T0174621

The
Global Economic
System

The Global Economic System

Iain Wallace

London and New York

First published 1990 by Unwin Hyman Ltd

Reprinted 1992, 1995
by Routledge
11 New Fetter Lane, London EC4P 4EE
29 West 35th Street, New York, NY 10001

Transferred to Digital Printing 2004

British Library Cataloguing in Publication Data
A catalogue record for this book is available from the British Library

Library of Congress Cataloguing in Publication Data
A catalogue record for this book is available from the Library of Congress

ISBN 0-415-08470-9

To
Barbara and Robert

Preface

Over ten years have elapsed since the idea of this book took shape. As the dynamism and unpredictability of the contemporary capitalist world-economy is one of its principal themes, there is a certain wry appropriateness to the delays the manuscript has suffered through intracorporate transfers and untimely takeovers in the transnational publishing industry. I am grateful to Unwin Hyman for 'adopting' the manuscript late in the day.

Fortunately, the book provides a perspective which I believe is even more pertinent today than when it was first conceived. Wherever one is situated, geographical analysis of economic change needs to start by 'taking the rest of the world seriously' (Thrift 1985). It must also incorporate the biophysical environment, and cultural and political context, of economic activity before it can successfully attempt to interpret currently formative institutions and processes. The philosophical stance from which I have approached this task is not expounded here, but it grows out of the position I developed in Wallace (1978).

I have been using the book, substantially in its present form, for a number of years in my second-year class at Carleton University. Others may lack the time or the inclination to cover all the ground in a single course, but should find sufficient material to build on in the chapters most suitable to their needs. I am encouraged by the number of students who have found the approach and content of this volume to have enriched their understanding of issues in subsequent and related courses, and who have made active use of it throughout their undergraduate careers. I trust others will find it similarly useful.

Acknowledgements

The bibliography is the best record of my intellectual indebtedness. For practical help, especially in the years before the advent of personal computers, I am glad to acknowledge the good-humoured and efficient support of secretaries in the Department of Geography at Carleton, especially Norma Lafrance, Mary Orser, and Judy Katz. Christine Earl prepared a number of the figures. For forbearance and all the other things that families contribute to an author, I am grateful to Katherine, Kristina, and Claire.

I am grateful to the following individuals and organizations who have kindly given permission for the reproduction of copyright material (figure numbers in parentheses):

De Vries, J. and reprinted by permission of Yale University Press Copyright © 1974 (2.3). Waddell, E.W. and reprinted by permission of University of Washington Press Copyright © 1972 (2.4). Hoffman R.C., W.N. Parker and E.L. Jones, and reprinted by permission of Princeton University Press Copyright © 1975 (2.6). Mutti, J and P. Morici and reprinted by permission of Richard D. Irwin Copyright © 1986 (3.3, 3.4). Taafe, E.J., R.L. Morrill and P.R. Gould with the permission of the American Geographical Society (3.6). Borchert, J.R. with the permission of the American Geographical Society (4.4 and 4.5). World Bank, Copyright © 1979, 1982, 1984 by the International Bank for Reconstruction and Development/The World Bank. Reprinted by permission of Oxford University Press, Inc. (3.11, 6.1, 11.1, 12.2, Tables 11.1 and 11.2). Smith, C.A. and reprinted by permission of Academic Press Copyright © 1976 (4.1, Table 4.1). Norton, R. D. and reprinted by permission of Academic Press Copyright © 1979. Kutcher, G.P. and P.L. Scandizzo and reprinted by permission of the Johns Hopkins University Press Copyright © 1982 (Table 11.3). Mager, N.H. and reprinted by permission of Media General Financial Servies Inc. Copyright © 1987 (1.1). Peet, R. (Table 11.4). Based on work by McLaughlin, J.F. and reprinted by permission of Harvard University Program on Information Resources Policy (6.5). Reprinted by permission of Pion Ltd Copyright

© 1988 (6.3). Latham, A.J.H. and reprinted by permission of Croom Helm Copyright © 1981 (4.10). Reprinted by permission of Times Books Ltd Copyright © 1982 (4.8 4.9). Roden, D., East, W.G., Rothwell, R., Malecki, E.J., Hart, K. and reprinted by permission of Cambridge University Press Copyright © 1936, 1973, 1980, 1982 (2.5,4.2,8.3,8.4, Table 11.5). Smith, W. and Field, N.C. and reprinted by permission of The Canadian Association of Geographers Copyright © 1968, 1984 (8.8 Table 9.1). Adapted from The Economist, 1978. (5.1). Bergmann, T. and reprinted by permission of Gower Publishing Copyright © 1975 (9.1, 9.4). Institute of Development Studies, University of Sussex 1978 (5.2). Investment Canada (6.2). Conzen, M.P. Reprinted by permission of Journal of Geography Copyright © 1983 (8.7). Micklin, P. 1988 (12.1). Denecke, D. and reprinted by permission of Ulster Folk and Transport Museum Copyright © 1976 (4.7). Harris, R.C. and reprinted by permission of the University of Wisconsin Press Copyright © 1966 (2.7). Vance, J.E. (3.5). Rogerson, C.M. and reprinted by permission South African Geographical Journal Copyright © 1980 (11.3). Brown, L.R. and reprinted by permission of Worldwatch Institute Copyright © 1982 (9.3). Wells, L.T. and reprinted by permission of Harvard University Copyright © 1972 (6.4). Johnston, R.J. and Taylor, P. J. and reprinted by permission of Basil Blackwell (3.10). Bairoch, P. and Cipolla, C.M. and reprinted by permission of Collins Copyright © 1973 (4.6). Hall, J.M., Townsend, A.R. and reprinted by permission of the Institute of British Geographers Copyright © 1970, 1987 (8.1, 8.5). Stern, R.M. and reprinted by permission of the University of Toronto Press Copyright © 1985 (Table 6.1). Eyre, S.R. (2.1,2.2). Browning, C.E and reprinted by permission of the University of Texas Press Copyright © 1981 by the Association of American Geographers (8.2). Rostow, W.W. and reprinted by permission of the University of Texas Press (10.1), Ray, D.M. and reproduced by permission of the Minister of Supply and Services Canada and Environment Canada (7.2). Taylor, D.R.F. (11.2). Datoo, B.A., Muller, P.O. and reprinted by permission of Economic Geography (3.9 Table 2.1).

Contents

List of tables

1 The global economy: orientations

Introduction

One of the most significant characteristics of the contemporary world is that the economic prospects of nations, regions, and even individual communities are fundamentally interrelated. Investment decisions made by foreign firms, the currency devaluation of a distant government, or the consequences of drought in another continent can all affect the livelihood of men and women thousands of kilometres away. In this sense at least, the concept of a *global* economic system is not hard to grasp.

For as long as people have traded goods between different regions, which takes us well back into prehistory, societies have experienced some degree of economic interdependence, and of the vulnerability to external events which it entails. But for centuries, the majority of the world's population was engaged in meeting its subsistence needs within localized, essentially self-sufficient societies (see Ch. 2). The evolution of an economy *based* upon the production and trade of marketed commodities was gradual. It began to emerge in Europe in the 16th century and reached fully global proportions by the early 20th century. Even the non-market-based *command economies* of the Soviet variety (see Ch. 9), which appeared after 1917, are increasingly linked into international markets. The speed and intensity of change within the global economic system have greatly increased since the early 1970s. As a result, it is more important than ever to understand the dynamics of economic interdependence and their impact in different parts of the world. That is essentially what this book sets out to explain.

The capitalist world-economy

Capitalism
The evolution of the global economic system has been interpreted by

Wallerstein (1976) as the emergence of the 'capitalist world-economy'. As Marx defined it, capitalism is a specific *mode of production*: a set of institutionalized practices within which human beings work to provide their material needs. The term is not narrowly 'economic', in the sense of referring only to mechanisms of resource allocation: it treats economic activity as an inherently social enterprise, governed by political and legal regulations which embody a particular set of cultural values. Historically, capitalism has flourished in societies which have embraced it as the economic basis of individual freedom. Capitalist societies are marked by a strong commitment to private property ownership, and by limited acceptance of government interference in the decisions of individuals to allocate resources (including their own labour) on the basis of the values which *markets* (impersonal mechanisms of exchange) set on them.

In comparison to other modes of production (reviewed in Ch. 2), capitalism is distinguished by three characteristics: its expansionary tendencies, its uneven results, and its enduring cultural appeal. The *expansionary tendencies* stem from the basic logic of a capitalist economy: the 'actors' (whether individuals or transnational corporations) constantly aim to accumulate capital derived from their business. One route to this goal is to expand the geographical area within which transactions are carried out. Another is to widen the sphere of social life within which markets operate (so that, for instance, care-giving moves out of the home and into commercial institutions). A third thrust is to promote scientific discovery and technological innovation as a means of harnessing new knowledge to commercial ends.

The *unevenness* of capitalist economic development is also multi-dimensional. Geographically, it is marked by the emergence of different-iated types of activity and different levels of prosperity in different areas, at global, national, and regional scales. How far these differences are 'disparities', *created* by the functioning of the capitalist economy (which many of them certainly are), and how far they are the undevised outcome of spatial variations in environmental conditions or contrasting cultural histories is a focus of theoretical disagreement (see Ch. 10). Nevertheless, the dynamics of economic and political interaction which sustain or erode these interregional inequalities are captured in the concept of '*core–periphery* relations', which Seers (1979a, p.xiii) argues is 'shorthand for a set of structural relationships' which link powerful core regions to weaker peripheries.

The temporal unevenness of capitalist development is evident in the *cycles* of growth, of varying duration and suggested cause, which economic historians have identified. Of particular significance for interpreting the recent evolution of the global economic system are the Kondratieff 'long waves', which have appeared since the Industrial Revolution (Fig. 1.1). Controversy surrounds their measurement and

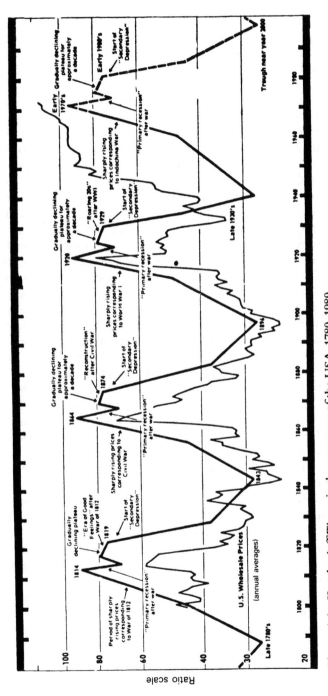

Figure 1.1 Kondratieff Waves in the economy of the USA, 1780–1980.
Source: Media General Financial Services Inc., Richmond, Va., 5 June 1974.

explanation, but the suggestion that they be regarded as 'distinct "modes of growth" of the total [social] system', in which decline is associated with 'a serious mismatch between the techno-economic subsystem and the socio-institutional framework' (Perez 1985, p. 36), indicates the complexity of forces producing them.

Marxist interpretations link the Kondratieff waves to the inherent dynamics of capitalist economies (Mandel 1980). Recession sets in as a natural consequence of the declining profitability of production. Overinvestment by competing firms, each attempting to capture larger markets during periods of expansion, and the success of wage-earners in increasing their share of gross revenues when business is buoyant eventually lead to excess (under-utilized) productive capacity and uncompetitive labour costs. Redundancies and bankruptcies mark a period of intensified competition, from which surviving firms emerge 'leaner and meaner', with a more profitable configuration of production establishments and labour costs to supply the prevailing level of market demand. It is crucial to recognize that this process does not proceed uniformly among firms, industrial sectors, regions, or national economies. The recovery of economic growth, which is by no means automatic, is associated with a new round of profit-enhancing innovations, embodied in industries whose locational preferences invariably differ from those of more traditional branches of production (see Ch. 8).

Emphasis on the revolutionary impact of technological innovation, as its effects spread throughout the economy, underlies the major non-Marxist explanations of long waves (Freeman 1987). The first Kondratieff expansion phase (approximately 1790–1815) was associated with the earliest technologies of the Industrial Revolution (mechanized textile production, improved iron production, and the steam engine). The fourth, following the Second World War, involved chemical, nuclear, and electronic technologies, together with rapid advances in aircraft performance. The fifth expansion, whose emergence Hall (1985) identifies, highlights the rôle of microelectronics (in such areas as computing, telecommunications, and industrial process control) and biotechnologies in transforming the economy.

The *cultural durability* of capitalism, despite the dehumanizing tendencies of unfettered market forces, has been explained in terms of its association, not just with freedom, but with 'progress' (Goudzwaard 1979). The predisposition towards growth and openness to change which characterizes all 'modern' societies finds broad and deep-seated acceptance; not least because it has promoted that technological mastery of the natural environment which underlies the high standard of living of the industrialized Western nations. To recognize this is not to deny the force of cultural and ethical critiques of capitalism, nor of sceptical appraisals of its economic 'benefits' (see Ch. 12). However, it should help us to appreciate why most contemporary state-socialist societies are facing identity crises arising out of internal demands for greater economic and political freedom, and for the greater affluence which is perceived to stem from it (see Ch. 9).

The world-economy

Wallerstein (1976) identifies the distinguishing characteristics of the capitalist world-economy as follows. There are two meaningful categories of socio-economic system. One consists of localized, self-sufficient groups which are *not* part of a hierarchical and geographically extensive system that demands some share of their human and material resources. Many 'traditional' societies were of this nature. The second category, which he terms 'world-systems', comprises societies which *are* geographically extensive and are specifically 'defined by the fact that their self-containment as an economic–material entity is based on extensive division of labor and that they contain within them a multiplicity of cultures' (1976, p. 230). There are two types of world-system; 'world-empires' and 'world-economies'. The Roman Empire is a good example of the former, in that it embodied a single political system which controlled, with varying degrees of effectiveness, the entire area within its boundaries. The prime example of a world-economy is the contemporary capitalist economic system. Capitalism's distinctiveness 'is based on the fact that the economic factors operate within an area *larger* than that which any political entity can totally control. This gives capitalists a freedom of maneuver that is structurally based' (Wallerstein 1976, p.230; italics added). The strategies and operations of modern transnational corporations (TNCs) demonstrate the reality of this freedom, which accounts for the tension frequently experienced in relations between them and territorially bounded national governments (see Ch. 7).

Within the capitalist world-economy, as in all world-systems, power is unequally distributed among its constituent units, whether these are defined as nations, social classes, or economic institutions. Between states, the inequalities are expressed in core-periphery relationships embracing political, economic, and military dimensions. A core state which enjoys overwhelming dominance within the world-economy is termed *hegemonic*, and for most of its history the capitalist world-economy has been shaped by the policies of a hegemonic state, most recently by Britain in the 19th century and the United States of America in the mid 20th. The common thrust of these policies has been towards the globalization of market transactions:

> Hegemony involves more than core status. It may be defined as a situation wherein the products of a given core state are produced

so efficiently that they are by and large competitive even in other core states, and therefore the given core state will be the primary beneficiary of a maximally free world market. Obviously, to take advantage of this productive superiority, such a state must be strong enough to prevent or minimize the erection of external and internal political barriers to the free flow of the factors of production (Wallerstein 1980, p.38).

The capitalist world-economy is in a turbulent era at the close of the 20th century. The post-1945 hegemony of the USA has gradually dissolved, and no single state looks capable of taking its place. Military power may still be overwhelmingly concentrated in the USA (and the USSR), but efficiency in manufacturing production and financial leadership are more obviously associated with Japan or West Germany. Moreover, core–periphery relationships at a global scale are much less clearly defined than they were in the 1960s, when the distinction between the First (industrialized capitalist) World and the Third World was coined. The emergence of 'high-income oil-exporting' nations, such as Saudi Arabia, and of 'newly industrializing countries' (NICs), such as Taiwan, not to mention the increasing size and diversity of the Chinese, Indian, Brazilian, and Mexican economies, makes the global economy a much more complex and dynamic system. Core states which feel vulnerable to the shifts in economic power, as the USA and many European nations have in recent years, are more prone to adopt *protectionist* trade policies, restricting imports, than to favour global free trade. The many Third World nations dependent upon international trade to finance their development are invariably the first casualties of such measures.

The changing structure of the economy

The decline of US hegemony, and of the relatively orderly conduct of international economic relations which was imposed by it, is clearly one source of turbulence in the contemporary global economy (Hirsch 1976b). The changing technological and institutional structure of economic activity is another. Recent and fundamentally novel developments have been summarized in terms of three divergencies:

(1) The primary-products economy has come 'uncoupled' from the industrial economy.
(2) In the industrial economy itself, production has come 'uncoupled' from employment.

(3) Capital movements rather than trade (in both goods and services) have become the driving force of the world economy (Drucker 1986, p.768).

We will review them in turn.

The rôle of resources
Despite scares in the 1970s that the world was running out of resources, popularized by the computer simulations of *The limits to growth* (Meadows *et al.* 1972) and given apparent support by the leap in raw material prices concurrent with the 1973 OPEC-induced oil supply crisis, it has become increasingly clear that most are in substantial oversupply relative to aggregate global demand (see p. 56). Between 1972 and 1985, world agricultural output rose by one third, with Africa the only continent not to share in the increase. By the mid-1980s, base-metal producers were receiving prices as low in real terms as those of the Depression in the 1930s. More recently, the chronic problems of subsidized agricultural overproduction in the European Community (EEC), the USA, and Japan have risen to the top of the agenda in international trade negotiations. It appears, in other words, that continued growth in the capitalist world–economy, albeit at a slower pace than during the 1950s and 1960s, requires proportionately fewer raw material inputs than in the past. Conservation, introduction of less resource-intensive technologies, and a structural shift in the industrialized nations from goods production towards services have all contributed to the 'uncoupling' of economic growth and the demand for resources.

Production and employment
In the period of sustained economic growth which followed the Second World War, the volume of world trade increased even faster than the rise in global output. Exchanges of manufactures between industrialized core nations grew particularly rapidly, and would have done so even more but for the proliferation of investment by TNCs, first from the USA and later from Europe, in foreign markets. Manufacturing investment in the larger developing countries was initially geared to import-substitution, typically resulting in uncompetitive products with limited markets. During the 1970s, however, a small group of Third World nations rapidly developed export markets for a widening range of manufactures. These NICs, spearheaded by the four Asian states of Hong Kong, Singapore, Taiwan, and (South) Korea, exploited changes in the world-economy to provide intense competition to labour-intensive industries in the core states of Western Europe and North America. Accelerated diffusion of sophisticated production technologies, improved transportation systems

involving containerization and air freight, and the greater managerial capacity for international coordination of production (whether by subsidiary enterprises or through subcontracting) permitted by advanced computer and telecommunication technologies made this development possible. The combination of effective government, an entrepreneurial culture, low wages, but high labour productivity was the basis on which the Asian NICs vigorously expanded their exports to core markets (see Ch. 7).

Within core nations, a major shift in the pattern of employment was already under way. Primary-sector employment, particularly in agriculture, continued to shrink in nations where it was not already down to about 5% of the working population. Manufacturing employment declined proportionately, but increased absolutely, until the early 1960s. Thereafter, in many industrialized countries, including the UK and the USA, it began an absolute decline *despite* continued growth in the volume of output (see Fig. 8.3). These shifts were not immediately apparent, and rising levels of industrial unemployment tended to be explained exclusively in terms of import penetration. Measures to limit the growth of manufactured imports from NICs, notably the Multi-Fibre Agreement governing trade in textiles (Steed 1981), became part of a growing tendency amongst core nations to protect domestic employment through non-tariff barriers (NTBs).

Job creation in industrialized countries during the 1960s and 1970s was overwhelmingly concentrated in the service sector, initially in such fields as health care, education, and public administration, and in a group of rapidly expanding 'producer services'. This latter group of activities is tangible evidence of the 'knowledge-intensity' of modern economies, but the nature and geographical distribution of the employment it generates is very different from that attached to material production. Moreover, as public-sector employment ceased to grow, and in many places declined in the 1980s, new jobs in the service sector became increasingly concentrated in insecure, part-time low-wage positions (see Ch. 8). As Drucker (1986, p.776) argues of the USA 'it is not the . . . economy that is being "de-industrialised", it is the . . . labor force'. Western European nations have been even less successful than the USA in limiting the proportion of their workforce without employment.

The primacy of finance
The third structural change in the world-economy has been the emergence since the early 1970s of what Drucker terms 'the "symbol" economy – capital movements, exchange rates and credit flows – as the flywheel of [economic activity], in place of the "real" economy – the flow of goods and services' (Drucker 1986, pp.781–2). Whereas, traditionally, international financial flows have seemed to be driven by trends in the

material economy of production and trade, it is now apparent that the largest movements of capital are essentially unrelated to shifts in the output and productivity of national economies (see p. 148). The greater instability of global financial markets since the convertibility of the US dollar into gold was abandoned in 1971 has been heightened by the greater willingness of governments to use movements in the exchange rate of their currency to gain comparative advantage in international trade. The relative cost of a country's factors of production, such as labour and natural resources, and of its manufactured outputs can now change rapidly and unpredictably, making long-term management of economic development or of social adjustment to trends in the international economic system much more difficult (if not impossible) to achieve.

These three shifts have involved profound changes in the geographical distribution of economic activity and the orientation of international transactions. They demonstrate the inability of individual sovereign states, even such residually powerful ones as the USA, to control the pace or trajectory of change within the capitalist world-economy. The profound internationalization of activity represented by the growth of global capital markets and corporate production systems severely limits the degree to which national governments can shelter their domestic economies and citizens from the impact of external developments. This puts a premium on the flexibility of national political and economic institutions to anticipate and accommodate change. The social and spatial implications of this pressure in Western core nations, which find it hard enough to adapt, are explored in Chapter 8. Most Third World states have infinitely less room to manoeuvre (see Ch. 10), especially those trapped in apparently permanent indebtedness to core institutions, notably banks and the International Monetary Fund (IMF).

Environment, culture and the economy

Values, ideologies, and economic development

During the period of sustained economic growth which followed the Second World War, public opinion in the industrialized nations of the West strongly supported growth, and readily accepted the wisdom of the 'experts' who 'ensured' it. Science and technology were held in high esteem (as pillars of the ideology of 'progress' as well as for their very tangible benefits), and technocratic 'planning' (in which economists played a privileged rôle) was viewed as rational, effective, and hence desirable. By the late 1960s, however, the impacts of large-scale projects to accommodate the demands of exponential economic growth (international airports, nuclear power stations, urban expressways, etc.) began to provoke increasingly vocal and determined opposition (Hall 1980). The values implicit in these developments were questioned, in particular for their apparent disregard of adverse environmental consequences and of social inequities in the distribution of costs and benefits. The experience of recession and unemployment in the early 1980s tested the depth of anti-growth sentiment and raised the question of how far it is an ideology of the affluent (who do not need new jobs). Yet a concern that high living standards should not come about at the expense of environmental degradation increasingly attracts popular support, and has received serious attention in business and government since the extent of potential global change has become apparent (see Ch. 12). Similar concerns have surfaced more recently in the industrialized state-socialist countries. Even in impoverished Third World nations, where the immediate challenge of survival often looms larger than the dangers of environmental deterioration, there is a growing awareness that economic growth has to be environmentally sustainable if it is to last. Throughout the world, therefore, technocratic interpretations of 'development' are being questioned.

A comparable but distinct reaction has emerged in many parts of the world against 'modernity'. The indifference, and frequent opposition, to traditional values of a culture based on commercial success, technological achievement, and 'progressive' social change has provoked varieties of opposition. The renewal of religious fundamentalism has been most noticeable in Islamic countries, but indigenous African religions and strands of American Protestantism have also had politically significant impacts. More generally, the policies and ideology of the liberal *welfare state* (see Ch. 5) have provoked in some countries (notably the UK and the USA in the 1980s) a conservative reaction. State intervention in the management of the economy and the provision of social welfare has been cut back, in the name of restoring freedom and initiative to individuals and the discipline of market forces to business. Similar priorities are imposed on indebted Third World governments by the international lending agencies. The result has been an increase in the economic, social, and geographical polarization which capitalism tends to generate when it is unregulated by representative governments. Yet, as we have already noted, the dynamics of the contemporary world-economy make it increasingly difficult for governments to resist international economic trends, even if they wish to (see p.273). Therefore, in economic and social affairs, as in environmental issues, the consistency and political legitimacy of State action is put in question (Habermas 1976). (The distinction between the "state", as a geo-political entity, and the "State", as a set of powerful institutions influencing social life, is elaborated in Ch. 5.)

The plan of this book

The interdependencies of the global economic system are 'global', both geographically and conceptually. An understanding of contemporary 'economic' issues may depend on some knowledge of environmental processes, cultural history, technology, or political ideologies. That is a tall order, and no single volume can cover all the ground! However, the following chapters do attempt to provide a framework within which detailed analysis can be anchored and its implications assessed.

Some of the basic relationships between a society and its environmental resource base are dealt with in Chapters 2 and 3, recognizing that different forms of social organization carry out material production in different ways. The economic and social evolution of the capitalist world-economy, a history which is critical to understanding many dimensions of contemporary economic development, is reviewed in Chapter 4. The following three chapters are focused on the two dominant institutions in the global economic system: the State (characteristics of which vary significantly around the world) and the transnational corporation, and the relationships between them (Ch. 7). The issues facing different groups of nations as they attempt to adjust to global economic change are reviewed in Chapters 8–11. Chapter 8 is focused on the industrialized Western nations, Chapter 9 on the industrialized state-socialist nations, and Chapters 10 and 11 on the development challenges facing Third World nations, both externally and domestically. The prospects for the global economic system are examined in the final chapter.

2   Society, the economy and the environment

Introduction

The emergence of the global economic system has presented
societies with a double challenge. Human beings have had to
develop knowledge of their natural environment and the resources
it provides. At the same time, they have had to devise forms
of social organization conducive to the successful management or
exploitation of these resources. Although the process of economic
development involves a progressively declining proportion of men
and women working directly with 'nature', in agriculture, fisheries,
and resource extraction, contemporary societies remain dependent
upon environmental processes for their basic needs, and their
standard of living is founded on technologies which harness the
energy/matter of the Universe. It is not surprising, therefore, that
the search for resources and the consequences of their exploitation
has significantly influenced the geographical evolution of the world-
economy.

The change in scale and complexity of social organization involved
in moving from essentially self-sufficient local communities to a
globally integrated economic system has brought about fundamental
transformations in the mode of production and in the distribution
of power. Growing specialization of economic activity and the need
to coordinate it over greater distances has been instrumental in the
emergence of impersonal market mechanisms and the evolution
of the 'modern' state (see Ch. 5). But these developments have
proceeded at a varying pace, under differing conditions, and with
diverse consequences in different parts of the globe, which accounts
for much of the diversity of wealth and power among contemporary
nations.

Natural resources

The environmental raw materials of economic activity are conventionally categorized as *renewable* (flow) or *non-renewable* (stock) resources. This distinction loosely corresponds to the time horizon within which natural processes replenish their supply, for both types are forms of low-entropy (usable) energy/matter. Renewable resources, such as trees, can generally be regenerated within a human lifespan, whereas stock resources, such as coal, accumulate only over geological time. In practice, the renewability of a 'renewable' resource depends upon societies allowing an adequate period during which low-entropy energy from the Sun can be absorbed (e.g. by photosynthesis) in forms directly available to human use. Unless a balance is maintained between the *rate* at which solar energy is captured in the natural environment and the *rate* at which the resultant resource is exploited, environmental degradation sets in and the 'renewable' resource becomes exhausted. This natural constraint on economic activity is most obvious in societies which lack the technological capacity to escape total dependence upon flow resources. The significance of the Industrial Revolution is that it marked the decisive breakthrough to an economy based on the exploitation of the added forms of environmental energy/matter stored in stock resources (see Ch. 4).

The consumption of environmental resources involves both quantitative changes in their availability *and* qualitative changes in the environment at large. The Second Law of Thermodynamics indicates that the utilization of low-entropy resources transforms them into high-entropy energy/matter, in which form they are unavailable for human purposes. For example, burning the 'free' (low-entropy) energy of a piece of coal produces the 'bound' (high-entropy) energy of heat, smoke, and ashes. And whereas the heat may contribute to economic output, the by-products usually constitute 'pollutants', changing the environment in ways which impose costs on society. The significance of this dimension of economic activity has gained recognition slowly, but it is now clearly established and has lead to remedial measures both at a local scale (e.g. air-quality controls) and globally (e.g. agreements to reduce atmospheric ozone depletion). More generally, all economic activity involves transformations of the natural environment, and the impacts generated by the contemporary world-economy are of such a scale that they enter increasingly into estimates of the viability of prevailing modes of economic development (World Commission on Environment and Development 1987).

Neither renewable nor non-renewable resources are distributed uniformly around the globe. The primary organic productivity of the world's ecosystems is largely determined by the amount of

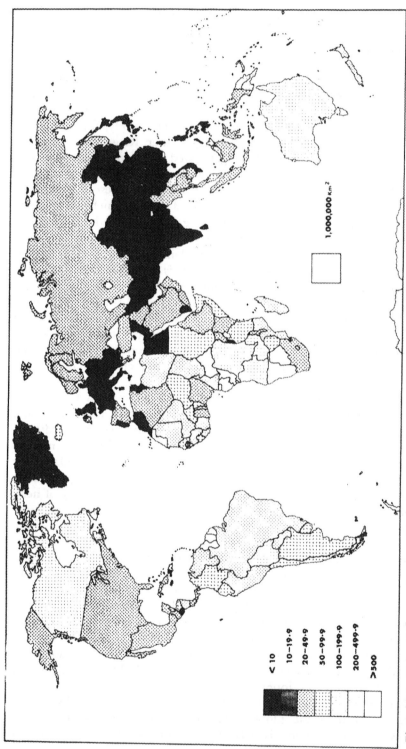

Figure 2.1 Potential per capita net primary organic productivity of the nations of the world (in tonnes).
Source: Eyre, S. R. 1978. *The real wealth of nations*. London: Edward Arnold; Figure 5.4.

< 10
10−19·9
20−49·9
50−99·9
100−199·9
200−499·9
>500

1,000,000 km²

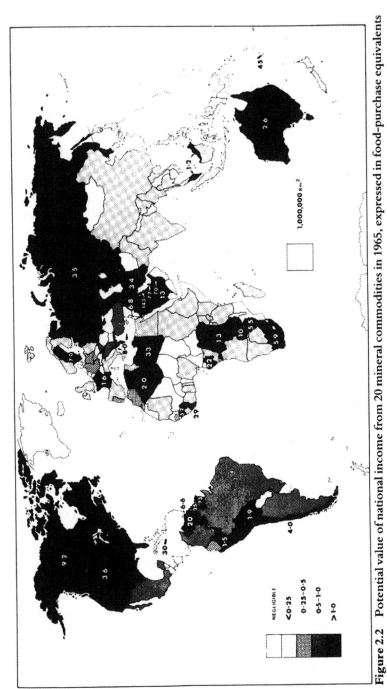

Figure 2.2 Potential value of national income from 20 mineral commodities in 1965, expressed in food-purchase equivalents (nutrition units per capita).

Source: Eyre, S. R. 1978 *The real wealth of nations.* London: Edward Arnold; Figure 7.1.

solar radiation and precipitation which they receive. As a result, there are major differences between world regions in the degree to which agricultural production can be intensified (see Table 2.1). In comparison to the wide geographical extent of major vegetation regions, economic deposits of stock resources tend to be much more localized, although some regions, such as southern Africa, are particularly well endowed with a variety of mineral deposits. Eyre (1978) has calculated for each nation an approximation of its 'real wealth' of flow (Fig. 2.1) and stock (Fig. 2.2) resources. Although former concepts of *environmental determinism* (which explained each society's development in terms of the environmental opportunities and constraints it faced) have been rightly discredited, some philosophies have gone to the other extreme and seriously underestimated the importance of environmental conditions for a nation's development prospects. This has been as true of Marxist planners of Soviet development (Matley 1966) as it has of Western economists working for the World Bank in the Third World (Kamarck 1976).

The economic significance of the geographical distribution of natural resources is increasingly evident in the interdependent global economy of the late 20th century. The concentration of the world's oil reserves in the Middle East, and the geopolitical leverage this has given to the Organization of Oil Exporting Countries (OPEC), is perhaps the most obvious example (see Ch. 7). More generally, the economic and environmental costs of energy-intensive processes in agriculture and industry are rising, especially of those dependent upon fossil fuels, so that technologies more reliant on the natural flow of solar energy have gained an advantage. For instance, oil-fired aluminium smelters in Europe and Japan have closed in recent years while hydroelectric-powered smelters in Canada have expanded.

Modes of production

Not only is economic activity unavoidably 'technological', in the sense that it involves the transformation of natural resources, but it is equally 'social', in that it involves the coordinated work of many people. We need to analyse how *individuals* make economic decisions (which has been the central concern of neoclassical economic theory), but we need also to examine the variety of ways in which *societies* may organize their production. This will enable us, among other things, to see more clearly what has been involved in the gradual emergence of the market-based capitalist world-economy as the dominant economic system.

The simplest form of social organization of economic activity is the household or localized community, which utilizes the resources of its immediate environment to achieve, effectively, self-sufficiency in material needs. The division of labour is weakly developed: individuals perform a variety of tasks, and insofar as there is an element of occupational specialization on the basis of age or sex, it is usually sanctioned by custom or religion. Exchanges of goods or services between members of the community take on the character of *reciprocal* obligations and express cultural values which are supportive of social *integration*. The social structure is relatively *egalitarian*, in that there is no permanent concentration of wealth and power in the hands of particular individuals or families (Marx defined this mode of production as 'primitive communism'). As the varied examples cited below indicate, any locally self-sufficient society, which is by definition based on harnessing flow resources at a sustainable rate, encounters basic environmental constraints on the emergence of social and economic inequality.

Societies organized on a wider geographical scale than that of a local community exhibit greater functional diversification. This implies a more pronounced division of labour and hence a greater need to exchange goods between individuals, often over long distances. How are these non-local transactions to be coordinated? Reciprocity cannot be sustained because of the complexity of the specific needs which communities require to have satisfied from distant sources. Moreover, a geographically extensive society is one in which internal variations in natural resource endowments invariably produce regional differences in the opportunities for creating wealth. Throughout history, this potential for the emergence of disparities has been exploited by powerful social groups to generate core-periphery relationships. At geographical scales at which reciprocity is an ineffective basis for coordinating the economy, transactions have been shaped either by *redistribution* within a hierarchical (non-egalitarian) society or by *market exchange*, which is the hallmark of a capitalist society (Polanyi 1968).

The social and economic stratification of redistributive societies, in which there is a systematic net flow of wealth from the mass of the population into the hands of an élite minority, presents the ruling élite with a problem. The concentration of social wealth has either to be culturally legitimized or else to be maintained by threat and force. Legitimation is a more efficient and peaceful mechanism, but élites have habitually used force in its absence. Prior to the emergence of the capitalist world-economy, the economic growth of Europe had proceeded for many centuries within a hierarchical society, the 'natural order' of which was sanctioned by the culturally dominant Church. The undermining of that world-view by the Renaissance and the Reformation laid the cultural foundation for the gradual acceptance of a society based on market exchange. But where the capitalist values of possessive individualism failed to elicit cultural support, within Europe or in the wider world, their penetration was accelerated with varying degrees of force (see Ch. 4).

The conceptual attraction of market exchange as a mechanism for social and economic integration is its implicit reliance on promises rather than threats. A *freely negotiated* exchange between *equals* (the underlying premise of perfect competition) represents a 'positive-sum game' (Boulding 1970). Each party to the transaction gains by entering into it, for its whole logic is that each *will* be better off as a result. In contrast, the use of force to *impose* an exchange (which, logically, cannot be between equals) is a much less productive form of social and economic relations. It imposes 'collection' costs on one party and forces the other to accept less than full value. These contrasted dynamics of market and forced exchange (which is a form of redistribution) are at the root of Wallerstein's distinction between 'world-economies' and 'world-empires'.

One may, of course, question whether the theoretical benefits of market exchange are realized in practice. Exchange in the modern economy routinely involves parties of unequal power as, for instance, when a large corporation does business with a small supplier; and governments in all developed capitalist countries have enacted laws to limit the abuse of market power by oligopolies (a small group of dominant firms). But, overall, market transactions are more satisfactory (to both parties) and more efficient as mechanisms of resource allocation than alternative types of exchange, and they allow societies to adapt relatively smoothly to changing economic conditions. The practical difficulties of organizing a flexible economy on the basis of centralized directive planning (see Ch. 9) have prompted a partial restoration of the market mechanism by both the Soviet and Chinese governments in recent years.

Nevertheless, the penetration of capitalism into societies based on another mode of production has invariably been socially disruptive. This is partly because market exchange became the basic organizing mechanism of *all* economic transactions, displacing traditional norms of exchange, but even more so because of the primacy of private property rights which it introduces. A cultural commitment to some form of *communal* property rights, especially with respect to land use, has characterized most 'pre-modern' societies, including that of medieval Europe under the feudal system (see p. 27). By extinguishing those provisions which allow all members of an agrarian society some form of access to land, and to the subsistence which is derived from it, the capitalist mode of production accelerates the emergence of a social class entirely dependent on wage labour to maintain its existence. Analysis of this development, the creation of a landless *proletariat* in a society increasingly dominated by property owners (first of agricultural land and then of industrial capital also), formed the substance of Marx's critique of capitalism and his vision of an economy based on socialized ownership of the means of production. The evolution of capitalist economies within the political framework of liberal democracy, and the expressions of capitalism associated with other political philosophies, are reviewed in Ch. 5 (see Fig. 5.2).

Population, resources and economic self-sufficiency

Perspectives on the population/ resources ratio

For most of human history, the volume of interregional and international trade has been strictly limited, and its impact on the everyday life of most people has been negligible. The basis of material welfare has therefore been the productivity of essentially self-sufficient local economies, harnessing renewable resources with simple technologies. Such societies have now been all but obliterated by the social and economic transformations which have brought them into the orbit of the world-economy. Nevertheless, the fundamental human dependence upon environmental resources is seen most clearly in these contexts. The dynamics of social adjustment to changing population/ resource ratios and the strategies for environmental management which they illustrate are of continuing relevance, despite the vastly increased technological capabilities and economic interdependencies of the modern world.

Self-sufficiency within a localized economy calls for sustainable and *diversified* resource exploitation. The need for diversity stems partly from the necessity of supplying all the community's basic needs, but it is also a risk-averting strategy. For example, the choice of agricultural crops will tend to include species with differing climatic tolerances, so as to ensure a harvest whatever the weather (Gould 1963). The lack of specialization in production does not mean that the economy is inefficient. Subsistence producers seek to minimize their work, but the benefits of efficiency are likely to be enjoyed in the form of increased leisure as much as in greater consumption. Indeed, Boserup (1965) argues that only under the pressure of a deteriorating population/ resource ratio, or a conscious desire to increase its level of material consumption, does a self-sufficient society commit itself to working harder.

The long-term viability of a self-sufficient economy depends on keeping the rate and manner of resource exploitation compatible with the reproductive capacity of local ecosystems. If a stable ecological equilibrium is disturbed, for whatever reason, the response tends to involve an increased rate of environmental exploitation to maintain or increase the living standards of the community. The degree to which this can be achieved *and sustained* within the locality depends partly on the potential productivity of the natural environment and partly on the society's readiness to adopt more intensive agricultural practices. Historically, the major causes of increased environmental exploitation have been one or more of the following: a steadily growing population; the invasion of a market economy, stimulating demand for tradeable surplus production; or the imposition of external rule (colonization), resulting in land expropriation or requiring production to pay taxes. Given the limited territorial extent of an economically self-sufficient society, the increased demand for resources presents it with two options. It may, if possible, gain access to additional territory by migration to a less densely populated region. Or it may attempt to replace its customary techniques of environmental exploitation with alternatives that raise the level of sustainable material output. But if neither is effective, the society will be forced to accept a progressive erosion of its living standards as the regenerative capacity of the natural environment is cumulatively undermined.

The evolution of the capitalist world-economy has involved all three outcomes (see Ch. 4). The 19th- and early-20th-century migrations from Europe to North America illustrated the first. The progressive intensification of European agriculture is an example of the second. And many regional famines (e.g. Ireland in the 1840s) have been witness to the third possibility.

Models of population–resource dynamics
Theoretical understanding of the dynamics of a changing population/ resource ratio has been advanced from two seemingly contradictory premises. Malthus, writing in 1798, argued that the Earth's resource potential is fixed, and that population growth is governed by that constraint. Boserup (1965) argued, conversely, that population growth is the constant factor, and the intensity of resource use will be adjusted to accommodate it. Both theories contain important insights (Grigg 1979).

Malthus published his *Essay on the principle of population* at a time when the population of England & Wales had increased by an unprecedented 50% in 50 years. He was concerned to demonstrate that, in the long run, a growing population is bound to exceed the carrying capacity of its resource base (i.e. its capacity to provide the means of subsistence). Malthus *assumed* that population has an inherent tendency to increase rapidly as long as there are adequate resources to support it, and also that a geometrical increase in population is accompanied by only an arithmetical increase in agricultural output. On this basis, it was not difficult to argue that, in the absence of checks on population growth, the eventual onset of famine was inevitable. The checks Malthus identified were *positive* (such as war and plague, which increase the death rate) or *preventative*, meaning social customs which tend to reduce the birth rate. The overall pessimism of his analysis has been echoed by all subsequent studies, such as *The limits to growth* (Meadows *et al.* 1972), which focus primarily on societies' ever-increasing demands on the globe's finite resources.

Malthus' conclusions can be faulted on a number of grounds, and Harvey (1974) has identified their ideological underpinnings. Anthropologists have shown that the assumption that population tends to expand out of control in technologically primitive societies is unfounded: many have maintained remarkably stable numbers over a long period, up to quite recent times (Wilkinson 1973). The assumption of limited prospects for increased agricultural output has also been undermined. Prior to Malthus' time, expanding the *area* under crops had been the only significant means of increasing the food supply in Europe. But even as he wrote there was accumulating evidence that *yields* could be raised by such measures as the new root-crop rotations, which were being introduced by progressive farmers in Britain and the Netherlands. Finally, Malthus did not foresee the impact that overseas migration from crowded Europe to the world's temperate grasslands would have on population/ resource ratios in the 19th century. On the other hand, the validity of Malthus' basic insight could be argued to be all the greater today because of the vastly increased (and still increasing) size of the global population, and the fact that the 'safety valve' of large-scale migration to 'empty' lands is no longer available.

Malthus argued that if a population grew to the point that the available resources could not provide a minimum subsistence, positive checks would inevitably cause its numbers to decline and restore a balance. Boserup (1965) disputed this, citing evidence that, even with low levels of technology, societies responded to a potential subsistence crisis by increasing the food-producing capability of their agricultural systems. Central to her theory is a trend towards increasing the *frequency* of land use as the demand for subsistence production rises. Instead of famine, Boserup saw an induced change in agricultural practice as the response to a worsening population/ resource ratio – *induced* because longer and more regular working hours are the price of maintaining an adequate food supply.

Links between these rival interpretations of population growth and its relationship to agricultural productivity can be demonstrated by a study of Figure 2.3. Let us initially assume a constant area of agricultural land and constant production technology. A small population (at D) is growing rapidly (the fertility rate is well above the mortality rate) and the increased application of labour to the land makes more efficient production possible. At E, population growth has continued and total agricultural output has been rising with it, but now the society begins to encounter *the law of diminishing returns*. There is a limit to the degree to which adding yet more labour can wring extra production out of a fixed supply of land and technology. Beyond E, where the marginal productivity of labour peaks, each additional member of the working population adds to total output but makes a contribution that is slightly less than that of the previous new worker. At B, output is the highest yet, but its rate of increase has slowed on account of the continuing decline in marginal productivity, which now equals average productivity. Continued population growth at this stage has the effect of reducing the living standards of everybody.

It is conceivable, as one of the fertility rate curves in Figure 2.3 suggests, that an awareness of declining marginal productivity will prompt the adoption of preventative checks to population growth, as the society takes steps to secure its continued prosperity. Stabilization of the population at B would allow for the highest sustainable average standard of living. Failure to curb population growth, however, implies gradual impoverishment, as average productivity progressively declines towards C. This process, whereby a society that presses against the carrying capacity of its resource base may be reduced towards a bare subsistence level, has been termed *agricultural involution* (Geertz 1963). Note, however, that whatever happens to the fertility rate, growing poverty and poorer diets come into play as positive checks on population growth, and raise the mortality rate. The Malthusian demographic cycle peaks at C, for beyond it the increased death rate brings about an absolute decline in population. The standard of living of the now smaller society will eventually improve (graphically, the output and productivity measures retrace their trajectories leftwards) to a point determined by the new balance that is achieved between births and deaths. Outmigration in response to the subsistence crisis would achieve a similar result.

Constant technology in Figure 2.3 is assumed. (It also assumes a homogeneous population. In an economically stratified society the poor will feel the consequences of declining productivity first.) Boserup argued, however, that the prospect of a Malthusian crisis is the stimulus to a *change* in technology, one that permits a higher level of sustainable output from a given resource base. This is achieved by increasing the intensity of land use and is expressed in the greater frequency of cropping or in a reduced period of fallow. The price of this greater productivity is a rising level of labour inputs to modify the environment of crop production, and hence it is generally associated with changes in the social structure of the population (Table 2.1). The range of options available to increase output will be governed by the climatic variables of length

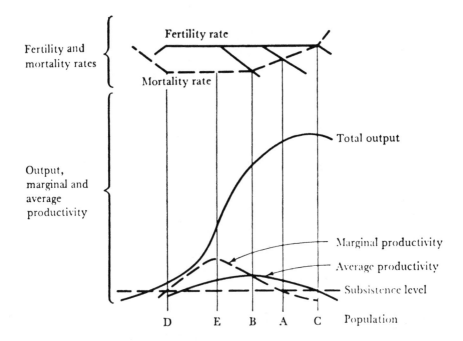

Figure 2.3 The link between demographic and economic conditions in a pre-industrial society.

D   Physiological minimum population.

E   Population level where the marginal productivity of labour is at a peak: i.e., total output grows at its fastest rate. Only after the economy passes this point can we expect an economically induced fall in the fertility rates.

B   Population level consonant with the highest per capita income level. If income is equally distributed, a fall in fertility rates could occur only after this point has been surpassed.

A   Population level at which "surplus" output (total output minus the subsistence requirements of the population) is greatest. An all-powerful ruler would find this the optimum population level. *Note:* The mortality and fertility curves associated with point A are not unique. We can claim only that a combination of positive and preventive checks on population and an autocratically optimum population level must fall between the the extremes of *B* and *C*.

C   Physiological maximum population.

The graph presents a static portrayal of the model; the units of the abcissa, population, should be understood as population at a given state of technology. Thus, if the economy adopts more productive techniques, the physical numbers of population are increased at each point, and the economy's position is pushed back (to the left).

Source: De Vries, J. 1974 *The Dutch rural economy in the golden age.* New Haven, Conn.: Yale University Press, Graph 3.1.

Table 2.1 Boserup's theory of agricultural change (modified after an unpublished paper by P. Porter).

	Agricultural stage				
	Forest fallow	Bush fallow	Short fallow	Annual cropping	Multi-cropping
Fallow systems					
duration of cropping	1–2 years	from 1–2 to 6–8 years	1–2 years	yearly	2–3 per year
fallow period	20–25 years	6–10 years	1–2 years	several months	negligible
fallow landscape	forest	bush	grass	none	none
Agricultural techniques					
agricultural tools	digging stick, axe, fire	hoe, axe, fire	hoe, plough, draught animals	mechanized equipment including tractors	mechanized equipment including tractors
fertilization methods	ash *in situ*	ash, supplemented by burnt or unburnt vegetable materials from outside	manure from droppings of draught animals	chemical fertilizers, green manuring, marling, composts, silt, etc.	chemical fertilizers, green manuring, marling, composts, silt, etc.
land improvements	none	none	rare	irrigation (in dry regions), terracing	irrigation, terracing
labour input	few hours of irregular work		pronounced seasonal peak		long hours of regular daily work

————— increasing amount per crop hectare —————⟶

Table 2.1 continued

	Agricultural stage					
	Forest fallow	Bush fallow	Short fallow	Annual cropping	Multi-cropping
Socio–economic structures						
land tenure	general right to cultivate land, no permanent occupation of plots	specific right to cultivate a given plot subject to authority	individual usufructory tenure, pledging to retain control	peasant ownership, small farms (freehold)	permanent ownership
			——— redistribution of land becomes less frequent ——→		
settlement form and transport network	unstable, dispersed settlements; trails	stable and larger settlements	permanent settlements; roads	urbanized to some extent	more intensive urbanization; more feeder roads
			——— increasing efficiency of road network ——→		
division of labour and exchange	rudimentary	some division of labour; village markets with part-time artisans	some professional artisans and traders	long work hours, greater division of labour, landless/ wage labourer group	long work hours, greater division of labour, outmigration
social and political organization	no central authority	a little more central authority	differentiated forms of social organization; domain over people	differentiated forms of social organization; domain over land	shift of power to (remote) urban centre

Source: adapted from Datoo, B.A. 1978. Toward a reformulation of Boserup's theory of agricultural change, *Economic Geography* **54**; Figure 1.

of growing season and amount of precipitation (for instance, multiple cropping of field crops is restricted to the Tropics). In terms of Figure 2.3, Boserup's theory implies that the shift to a production system of greater intensity both increases the number of people represented at each point between D and C, and shifts the productivity and output performance of the society leftwards.

The history of agricultural development around the world confirms the validity of Boserup's fundamental argument – sinking into a Malthusian crisis is not the inevitable fate of a growing population. The changed practices which a society accepts are not 'an overnight switch to an alien system with a novel tool kit, but a gradual alteration of the total effort devoted to somewhat more intensive production' (Spooner & Netting, quoted by Datoo 1978, p.139). As a result, it is not surprising that most agrarian societies have practised differing levels of land-use intensity side by side, as do the Enga (see below). Nevertheless, as Table 2.1 indicates, there are limits to the potential for more intensified production from any given tract of land. Regional subsistence crises have been recurrent events in pre-industrial societies, and they continue to occur in a world which has the technological capability significantly to mitigate their consequences. Moreover, a crude increase in the population/ resource ratio is not the only land-use pressure which societies have to contend with. As noted above, the need to earn a cash income, whether to pay taxes or to purchase non-local goods, necessarily involves the devotion of some land to cash crops at the expense of staple foods. It is premature, therefore, to argue that Malthus' pessimistic conclusions are totally invalidated by the route to increased productivity outlined by Boserup. Even if the Earth as a whole can feed a larger population than the current 5.5 billion, agricultural involution is likely to continue to occur in particularly stressed environments.

Self-sufficient economies
The following four case studies demonstrate specific environmental challenges faced by approximately self-sufficient societies and the significance of social relationships in determining the precise form of response. The Enga of New Guinea represent indigenous tropical agriculturalists with a very low level of technological capability. The common-field village economy of feudal Europe was the agrarian setting for the emergence of that continent's early capitalist economy. The self-sufficient rural economies developed by Europeans overseas in the 17th and 18th centuries illustrate the significance of contrasted population/ resource ratios. The rural economy of China in the 1960s demonstrates the links between social and environmental transformation. In every case, the harnessing of environmental resources and the efficient organisation of space go hand in hand,

with results that reflect the demographic and technological capacities of each society.

The Raiapu Enga   The Enga are a group of 30 000 people in the highlands of New Guinea, living in an environment that is marginal for the successful cultivation of their dietary staple, the sweet potato (Waddell 1972). To counter the threat of frost damage, the Enga plant this crop on top of specially constructed mounds. In addition, they farm other land at a lower level of intensity, growing a variety of vegetables in 'mixed gardens' formed by shifting cultivation in the surrounding forest (Fig. 2.4). The third component of their subsistence agriculture is pig-rearing, which provides both meat and a store of wealth within their social system. Overall, the Enga economy displays 'a remarkably sophisticated and stable adaptation [to its natural environment] with a simple technology and under high population densities' (Waddell 1972, p. 6).

The spatial organization of Engan agriculture demonstrates efficient solutions to two problems. Sweet potatoes are basic to the diet of the pigs and the human population alike: it is therefore necessary to regulate the pigs' access to the potato plots! The layout of farmsteads is designed so as to minimize the need for fencing, and to ensure that adjacent holdings maintain the integrity of the major fence line which separates the potatoes from the forest and 'mixed garden' zone where the pigs roam. Secondly, the greater amount of labour required for potato cultivation as compared to the 'mixed gardens' would suggest that dwellings be located close to the plots. In fact, they are situated at approximately the point of minimum aggregate travel for the cultivator.

At the time of Waddell's study (*c.* 1970), the self-sufficient economy of the Enga was being undermined by two sorts of external influence. The arrival of agencies promoting modern standards of hygiene and health care had reduced infant mortality, destroying the previous relative stability of the population/ resource ratio. Growing population pressure was resulting in shorter fallows in the 'mixed garden' zone, reducing its long-term fertility and requiring increased labour inputs (for weeding) during cultivation. Simultaneously, the desire to acquire commodities that were increasingly available from the outside world was prompting the Enga to devote land and effort to cash crop production at the expense of their staple sweet potato. Pigmeat, too, was becoming prized for its market, rather than its ceremonial, value. The decay of traditional social structures and their replacement by privatized, commercially oriented farm households was clearly under way.

Common-field villages of feudal Europe   Feudalism has been interpreted as a form of social and economic organization that emerged in response to

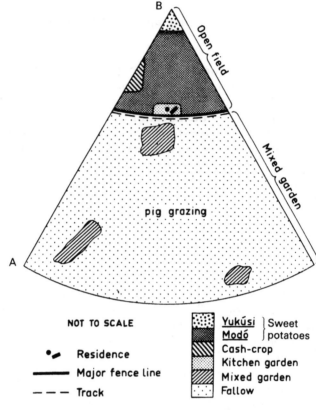

NOT TO SCALE

•⌣ Residence

——— Major fence line

– – – Track

Yukúsi ⎱ Sweet
Modó ⎰ potatoes
Cash-crop
Kitchen garden
Mixed garden
Fallow

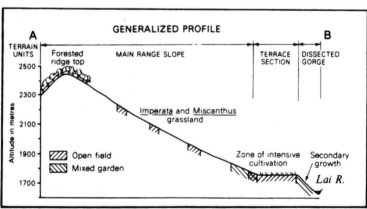

Figure 2.4 The land-use pattern of an Engan farmstead.
Sources: Waddell, E.W. 1972. *The Mound builders: Agricultural Practices, Environment and Society in the Central Highlands of New Guinea*. Seattle: University of Washington Press; Figure 20; Brookfield, H. (ed.) 1973. *The Pacific in Transition*; London: Edward Arnold, Figure 2.1.

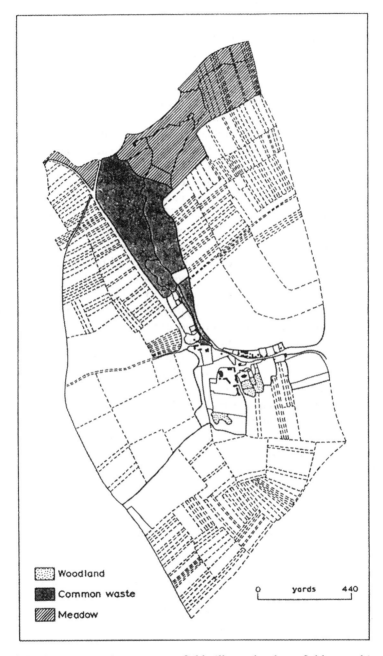

Figure 2.5 A representative common field village: the three-field township of Cuxham, Oxfordshire, in 1767.

Source: Roden, D. 1973. Field systems of the Chiltern Hills and their environs, in *Studies of field systems in the British Isles*, A. R. H. Baker & R. A. Butlin, (eds)., Cambridge: Cambridge University Press; Figure 8.4.

the period of anarchy which followed the collapse of the Roman Empire. Peace was restored through a complex political process, resulting in the creation of a hierarchical society held together by reciprocal obligations. Territory over which a ruler established his authority was distributed among lesser nobles, who were granted rights to use it in exchange for rendering military and administrative services. Lesser nobles (lords), in turn, granted peasants land-use rights in exchange for services and payments in kind. The subsistence economy of each village thus helped to maintain a wider society of culturally legitimated authorities (the nobility and the Church), whose local representatives mediated most of the limited trade which took place.

Common-field villages originated in the 8th century in response to two resource constraints. In long-settled areas, the earlier economy of extensive pastoralism and mixed farming, in which peasant households enjoyed considerable independence, was eroded by a gradually increasing population which whittled away areas of (uncolonized) wasteland. These had previously been a free good to the peasantry and the home of their livestock, but the intensifying competition for land increased the power of the lords to limit and exploit peasant production. This goal was more rapidly accomplished by concentrating the population in nucleated settlements. Simultaneously, the shrinking capacity of the waste to support livestock (essential for food, raw materials and as draft animals) stimulated an agrarian innovation to combine arable production and stock-rearing on the same land. This required a greater use of fencing and an agreed method of controlling potential conflicts in the use of arable land. The prevailing culture, based on a corporate understanding of society, fostered a communal rather than an individualistic solution. And the limitations of the prevailing technologies of fencing and ploughing favoured the specific spatial solution that emerged (Fig. 2.5).

The land of a typical village was divided into three large fields to minimize the need for fences. Land use rotated annually, with one field always in fallow and available to pasture livestock. Within each arable field, land was divided into many long, thin strips, and each family had cultivation rights over a number of such strips scattered throughout the village. This distinctive pattern of land use minimized the effort involved in ploughing, for turning was difficult. More crucially, perhaps, it minimized the risk of crop failure: the (inefficient) scattering of strips has been interpreted as the premium 'on an insurance policy in a milieu in which agricultural yields were low and unpredictable' and in which the costs of a poor harvest were high (McCloskey 1975a, p. 115).

Some of the broad environmental and social factors which conditioned the spread of common field villages can be inferred from Fig. 2.6. In particular, upland areas tended neither to favour staple grain

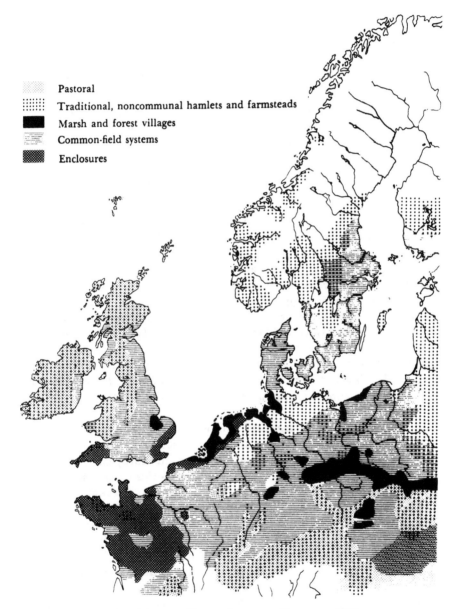

Figure 2.6 Agricultural 'systems' in northwestern Europe c. 1300.
Source: Hoffman, R.C. 1975. Medieval origins of the common fields. In *European Peasants and their Markets: Essays in Agrarian Economic History*, W.N. Parker & E.L. Jones (eds)., Princeton NJ: Princeton University Press; Figure 1.1.

production climatically nor did their sparser settlement promise lords adequate returns for the effort of imposing the common-field system. In parts of central Europe, where feudal society dissolved slowly, the essential self-sufficiency of the village economy lasted for almost 1000 years, until demographic pressure on a finite resource base and in a production system which hampered further improvements in productivity resulted in a subsistence crisis and outmigration (see p. 87). In northwestern Europe, however, and particularly in England, the early development of a market-oriented economy and cultural acceptance of land *alienation* (rights of private ownership) promoted the dissolution of feudal social relations and the common-field village economy from as early as the 14th century.

In a conservative society, the communal character of common-field agriculture was an obstacle to innovation. Hence land alienation and the *enclosure* of private farmsteads out of the stock of formerly common land was a necessary step in the pursuit of higher productivity. Individuals could thereby better ensure that they reaped the economic benefits of agricultural improvements such as breeding superior livestock or the introduction of new crops (North & Thomas 1973). Growing urban markets for food were an incentive to such development. In terms of Table 2.1, enclosure did not necessarily involve a switch to more intensive land use (indeed the reverse if wool was more profitable than grain), but by the time of *the enclosure movement* in the mid 18th century (see p. 73), England's growing population stimulated the switch from short fallow to annual cropping, based on new crop rotations.

Early European agriculture overseas   This is not a pure example of localized self-sufficiency, for the growth of transplanted societies was aided by links with their homelands. However, not all European settlements evolved along the lines of England's American colonies, which were characterized from the very beginning by external trade and a market-oriented domestic economy. French settlers in Canada, and Dutch in South Africa, found themselves remote from centres of power and commercial activity, and confronted by a physical environment markedly different from that they had known in the Old World. This combination of factors promoted the emergence of societies distinctively different from their European roots, despite their remaining nominally subordinate to the same social and political hierarchies. Greater independence and a higher standard of living reflected the lower population/ resource ratios of the colonial setting (Harris & Guelke 1977).

Despite a superficial resemblance between the strip cultivation of common fields in feudal Europe and the strip pattern of land holdings adopted by French settlers in the lower St Lawrence Valley (Fig. 2.7), the

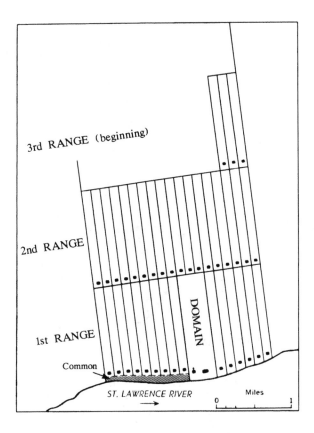

Figure 2.7 Agricultural settlement in a hypothetical *seigneurie*, St Lawrence Valley.
Source: compiled from Harris, R.C. 1966. *The Seigneurial System in Early Canada: A Geographical Study*. Madison: University of Wisconsin Press; Figures 7–1 and 9–1.

agrarian economy of the two regions was very different. In France, the social control of the noble was effective because land was comparatively scarce and labour comparatively plentiful. In Canada, the opposite was true; the imposition of excessive burdens on the peasant (*habitant*) tended to prompt migration of the household to the territory of a more congenial lord (*seigneur*) who was anxious to populate his territory and thereby cover his expenses. Despite calls from France for the creation of nucleated settlements to tighten political control, only six were established in 140 years of French rule. Similarly, Dutch settlers who moved inland from the Cape in South Africa found an extensive frontier in the dry, stock-farming regions of the interior, which gave them ample scope for establishing self-sufficient holdings subject to minimal imposition by the authorities.

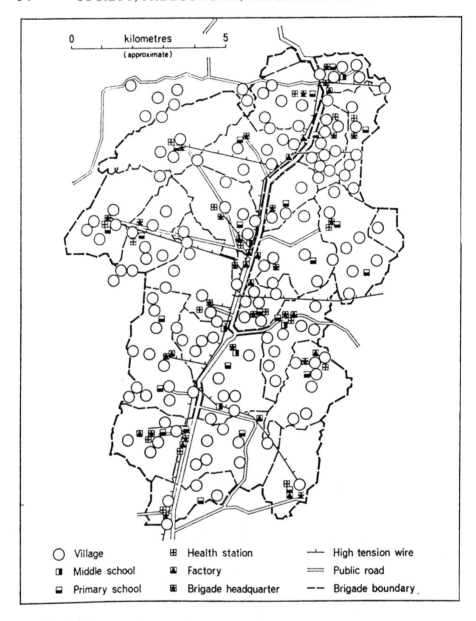

	Village	⊞	Health station	—ⵏ—	High tension wire
◖	Middle school	◪	Factory	══	Public road
▱	Primary school	⊞	Brigade headquarter	---	Brigade boundary

Figure 2.8a and b A representative commune, southern China, 1966.
Source: Buchanan, K. 1970. *The transformation of the Chinese Earth*. London: G. Bell and Sons; Figures 48 and 49.

These two settler societies developed a more egalitarian and homo-geneous social order than that of their homelands, primarily because they introduced European cultural perspectives on the family and land into

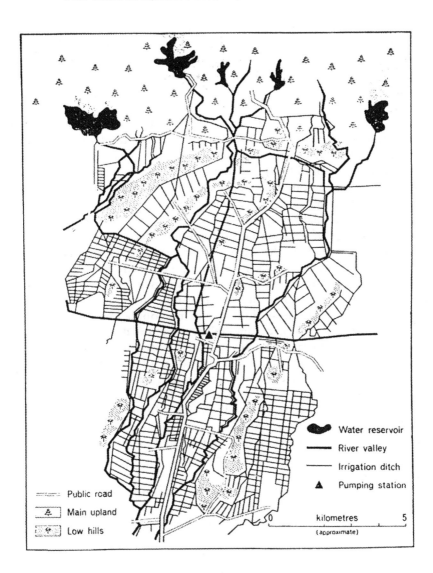

Water reservoir

River valley

Irrigation ditch

▲ Pumping station

Public road

Main upland

Low hills

kilometres   5
0
(approximate)

Figure 2.8b

settings where land was cheap and markets for farm produce were poor. Because land was readily available, only the very poorest were denied the prospect of carving out a property of their own. But as long as markets were poor, it was impossible to grow rich by one's work on, or ownership of, land. Hence self-sufficient farm households, enjoying a comfortable standard of living, were the prevailing unit of society. In neither Canada nor South Africa (except around Cape Town) were there urban markets of any size, nor did agricultural commodities become competitive in European markets until the late 18th century. These societies were eventually transformed by the spread of commercialization and, in the St Lawrence Valley, the build-up of demographic pressure leading to outmigration.

Rural China in the 1960s   At this time, rural China was embroiled in the radical agrarian change which had been initiated by the communist revolution of 1949. Despite the key rôle of non-local influences (both material and sociopolitical) and the selective introduction of modern technologies into an otherwise traditional agriculture, approximately the same proportion (80%) of rural production was consumed locally as had been typical of feudal Europe.

The challenge of agrarian tansformation under communist rule can only be appreciated in the light of the prior conditions (Buchanan 1970). For a century, a weak central government had allowed the land-owning class to exert growing power over the economy at the local level. This aggravated the poverty and helplessness of individual peasant families which, in addition to a rising population/ resource ratio, stemmed from their historic vulnerability to natural disasters and disease. Over much of northern China, low and highly variable rainfall produced recurrent droughts and famines. Large areas of central and southern China were plagued by equally unpredictable excess rainfall, resulting in widespread and protracted flooding. Centuries of deforestation had increased the associated soil erosion. Giving the rural masses much greater control over their natural environment was therefore a cornerstone of communist policy, but it could not be achieved without a corresponding increase in their control over their social and economic environment.

Following the initial dispossession of landlords and rich peasants, many of whom were killed, the productive resources of rural society were placed in the hands of those who directly used them. Ideological and pragmatic considerations prompted a rapid move away from agrarian organization based on household units towards communal institutions of increasing size, until the early *communes* were found to be unmanageably large and were subdivided. A commune in southern China, as it existed in 1966, is illustrated in Figure 2.8. A population of 50000, living on an area of 6000 ha, irrigated 98% of its arable land, which was devoted predominantly to double-cropped rice. This intensive agriculture was possible in a region of formerly disastrous rainfall fluctuations only because of the large-scale mobilization of peasant labour to modify the natural environment. Control of water resources over the entire catchment area involved building storage reservoirs in the hills, re-aligning small river channels, and stabilizing hill slopes by afforestation. The limited but strategic use of industrial inputs from outside the commune took the form of electricity, a pumping station, and small quantities of chemical fertilizer to augment organic manure (Fig. 2.8b). But the commune was more than just an agricultural production unit: it was an organization for coordinating the whole range of rural life, including the provision of educational, health, and local administrative services (Fig. 2.8a).

Conclusion

The creation of a stable and adequate level of local self-sufficiency was the first and mammoth task facing the Chinese communist regime. The speed of agrarian change and its links, however limited in most regions, with a growing industrial base, differentiate this society from the others reviewed above. So too does the world-view within which an increase in agricultural productivity played a consciously assigned rôle in the broader project of national economic development. Nevertheless, there is continuity among these case studies in the challenge facing a population to secure its basic subsistence within defined environmental parameters. There is contemporary relevance also in the reminder that the achievement of economic development and social wellbeing on any appreciable scale calls for careful attention to preserving and enhancing the productivity of natural environmental systems. This frequently requires social and political transformation to permit desired changes in resource-use practices (Blaikie 1985).

3 *Population and resources in an industrialized world-economy*

Introduction

The growth of specialized regional economies, in which a rising proportion of the population devotes some if not all of their resources to the production of marketable commodities, was well under way before the Industrial Revolution (see Ch. 4). The late 18th century nevertheless marked a turning point in the development of the capitalist world-economy, one that reflected two fundamental changes in the population/ resource ratio. The first of these was the decisive, although by no means overnight, transition from an economy fundamentally dependent upon renewable resources and energy supplies to one based on fossil fuels and stock resources. This quantum leap in the availability of metals and mechanical energy permitted the more *intensive* exploitation of locally available resources (renewable and non-renewable) and, by revolutionizing transportation technology, the tapping of resources over a much more *extensive* geographical area.

The second development was the onset of the *demographic transition* (Fig. 3.1). After centuries of very gradual growth, the world's population began to increase exponentially. Initially confined to countries in northwestern Europe (Malthus was one of the first to comment on it), the acceleration spread during the 19th century to the rest of Europe and to areas of European settlement overseas. (The transition did not begin to appear elsewhere until 1920 and only after 1950 did it really take root in the Third World (Chung 1970). Its causes and timing are analysed in their historical context in Chapter 4.) The dramatic population increase in the European core of the capitalist world-economy, combined with the associated growth in demand for

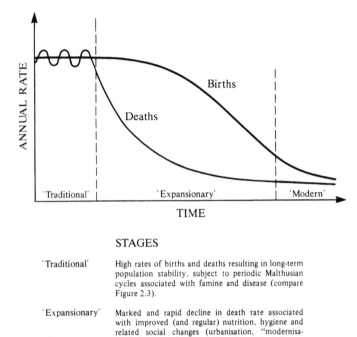

TIME

STAGES

'Traditional'   High rates of births and deaths resulting in long-term
                    population stability, subject to periodic Malthusian
                    cycles associated with famine and disease (compare
                    Figure 2.3).

'Expansionary'   Marked and rapid decline in death rate associated
                    with improved (and regular) nutrition, hygiene and
                    related social changes (urbanisation, "modernisa-
                    tion"). Reduced birth rate also reflects social change,
                    but lags by a (varying) number of generations, result-
                    ing in rapid population growth.

'Modern'   Low rates of births (major role of contraceptives) and
                    deaths resulting in a relatively stable population. Slow
                    growth the norm, but some nations may experience
                    absolute decline from current levels.

Figure 3.1   The demographic transition.

resources, were powerful stimulants to the maturation of an economy of regional specialization and long-distance trade.

Superficially, the expanded range of available resources at the heart of the Industrial Revolution and the increased population resulting from the demographic transition might simply have counterbalanced each other. The actual relationship between the two developments has been much more complex. The essence of the demographic transition as a model of population dynamics is that growth rates, having risen, subsequently decline. Moreover, the timing and speed of decline appear to be closely associated with those social and economic changes brought about by industrialization. Certainly, the experience of the global core has been that resource consumption has outpaced population growth and thereby raised material standards of living. The crucial question may therefore be to ask why the relationship has been weaker or different in much of the Third World, and that is addressed in Chapter

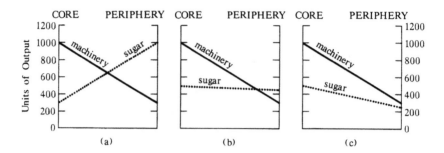

Figure 3.2 The basis of comparative advantage.
Ricardo's formulation involves two products and a single resource input, labour. It assumes the cost of inter-regional trade is zero. In each case, the core enjoys an absolute advantage in machinery production. In (b) and (c) it also has an absolute advantage in sugar production. The periphery enjoys an absolute advantage in sugar production only in (a), yet its comparative advantage in sugar production in all three cases can be demonstrated from examination of the sugar/machinery exchange ratios:

(a) periphery trades 1.0 units of sugar for 3.3 units of machinery from the core, foregoing the 0.3 units of machinery its labour could have produced.
(b) periphery trades 1.0 units of sugar for 2.0 units of machinery from the core, foregoing the 0.67 units of machinery its labour could have produced.
(c) periphery trades 1.0 units of sugar for 2.0 units of machinery from the core, foregoing the 1.2 units of machinery its labour could have produced.

The comparative advantage of the core in machinery production can be calculated similary.

10. Our present objective is to explore the geographical characteristics of resource use in an industrialized and increasingly internationalized economy.

Resource endowments and the geography of trade

Comparative advantage
Theoretical analysis of the benefits to be derived from specialized production and international trade was central to the work of the classical economists. Ricardo (1817), in particular, argued that if each nation were to specialize in producing the commodity for which it enjoyed a *comparative advantage*, it could increase its aggregate wealth, even if a trading partner enjoyed an *absolute* advantage in the production of the same commodity (Fig. 3.2). Underlying this formal analysis is the concept of *opportunity cost*; that the basis for calculating the optimum use to which labour or resources could be put is to estimate the

returns which would be *foregone* if they were deployed in an alternative manner. The expanded neoclassical formulation of international trade theory is the Hecksher–Ohlin model. This recognizes that opportunity costs increase as the degree of specialization increases; so it is never total. Specialization reflects each nation's *mix* of productive resources: its exported goods will embody much of those factors which are domestically abundant and little of those which are scarce. Its imports will have the opposite characteristics. Contemporary patterns of international trade validate the general thrust of the Hecksher–Ohlin theory, in that export/ import ratios are in line with variations in factor endowment (Figs 3.3 & 4). However, its adequacy can be questioned on both theoretical and empirical grounds (see p. 225). The latter include the fact that increasing proportions of global trade take place between countries with high and similar incomes, and that they consist of similar manufactured products (Lindert 1986).

Theoretical dissatisfaction with the Hecksher-Ohlin model stems from the general conceptual limitations of neoclassical economics. By assuming a static and perfectly competitive environment, it conceals the fact that international trade takes place within a dynamic set of core-periphery relations. Nineteenth-century writers recognized the force of comparative advantage arguments for perpetuating the core's dominance in specialized manufacturing and the periphery's restriction to primary commodity exports (Chisholm 1982). In the short run, such a division of labour does maximize the efficiency of resource use, but it is not a recipe for economic *development* in the periphery. Significantly, the exchange of British manufactures for North American resources in the 18th century conformed to the principles of comparative advantage. But, as soon as the USA and Canada became independent nations, their governments recognized the desirability of changing their relative factor endowment by stimulating domestic industrial activity, albeit at the cost of tariff protection which raised the price of manufactured goods (see Ch. 4). Precisely the same considerations face peripheral nations in the contemporary global economy (see Ch. 11). The speed and extent to which a state can change the basis of its comparative advantage at a politically and economically acceptable cost is a complex question, which can only be meaningfully addressed in a specific context.

Other assumptions of the Hecksher–Ohlin model which invite critique include the following: that all nations have access to the same choice of production technologies; that productive resources are fully employed in each nation and are not internationally transferable; that factors of production can be switched with ease between different sectors of each domestic economy in a perfectly competitive environment; and that neither governments nor powerful TNCs have any influence on the flows of internationally traded commodities (Todaro 1977). In fact,

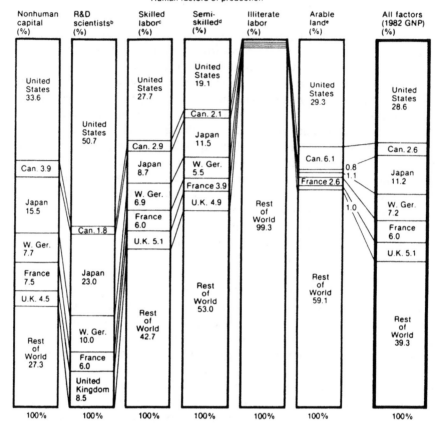

Figure 3.3 Shares of the 'world's'[a] factor endowments, 1980.

Estimates for individual factors derived from Mutti, J. and Morici, P. 1983. *Changing patterns of US industrial activity and comparative advantage*, Washington, DC: National Planning Association. Note: All estimates are rough approximations.

[a]Here the "world" GNP excludes the the Soviet Union, Eastern Europe, and Kampuchea for want of comparable data. For the individual factors, it also excludes other small countries acounting for less than 15 percent of market economy gross product.

[b]From data provided by the U.S. National Science Foundation. It is an exaggeration to show, as here, that these six countries had 100 percent of the R&D scientists. Had data been available for other countries, however, the six countries would still have accounted for at least 85 percent of the "world" total.

[c]Workers in professional and technical categories (ILO data).

[d]All literate workers who are not professional-technical.

[e]Based on land area measurements adjusted for different productivity in different climatic zones, for 1975.

Source: Lindert, P.H. 1986 *International Economics*, 8 Ed, Homewood, Ill.: Irwin; Figure 2.5.

core and peripheral states are significantly differentiated with respect to each one of these issues. As a particular example, over one third of the volume of US trade in manufactures consists of shipments between US TNCs and their foreign affiliates (McConnell 1986), transactions usually involving *transfer pricing* (see p.147).

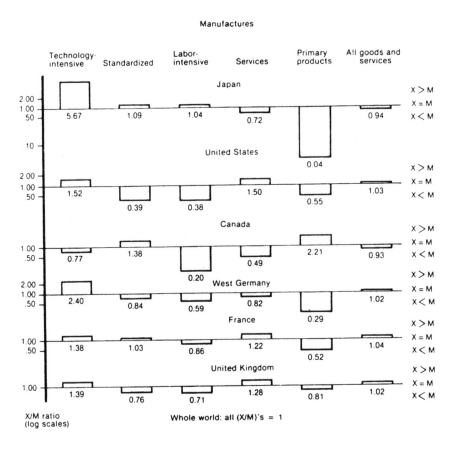

Figure 3.4 Patterns of comparative advantage: export/import (X/M) ratios in six leading countries, 1979.
Calculations derived from Mutti, J. and Morici, P. 1983. *Changing patterns of US industrial activity and comparative advantage*, Washington, DC: National Planning Association and IMF data. Technology-intensive manufactures=transport equipment (autos, planes, ships, motorbikes etc.) machinery, chemicals and professional goods. Labour-intensive manufactures=textiles, apparel, footwear and leather goods. Standardized manufactures = all manufactures *less* technology-intensive and labour-intensive manufactures.
Source: Lindert, P.H. 1986. *International Economics*, 8 Ed, Homewood Ill.: Irwin; Figure 2.6.

Export staple economies

Reflection on the evolution of peripheral economies shaped by European penetration has produced other theoretical frameworks for analysing resource development and trade. Innis, whose *staple theory* draws on Canadian experience (Watkins 1963), and North (1955), whose similar *export base* theory reflects US experience, both focus on the low population/ resource ratio and relatively open and egalitarian social conditions associated with '*colonies of settlement*'. Commercial production of exportable commodities originating from the land (or sea in the case of eastern Northern America, but significantly not from mines in the early stages of settlement) involved the widespread participation of family units in the process of capital formation which we associate with the spatially extensive process of 'pushing back the frontier'. The entry of such territories as peripheral suppliers to core European markets, and their subsequent evolution in (or from) that rôle, provides a strong geographical component to theories of resource-based regional development.

Innis discerned that the economy and human geography of an 'empty' peripheral nation are shaped by characteristics of the natural resource *staple* on which its trade depends. As the local bridgehead of the capitalist world-economy, the industry which exploits the staple resource is the *leading* sector of the domestic economy. Demand for the staple in core nations promotes immigration and initial capital investment. The size and prosperity of the peripheral economy depends wholly, at this stage, on the success of the staple. Subsequent economic development depends on the degree to which this leading sector stimulates the emergence of other forms of production. One obvious demand is for a transportation system which will make the staple product accessible to world markets. Another is for goods, including capital equipment, used by the staple industry. These demands represent *backward linkages* generated by the expanding resource sector. As the peripheral economy grows in size, it becomes increasingly profitable to supply these goods domestically rather than to import them from the core. The success of the staple industry will also create opportunities for *forward linkages*, in the form of manufacturing industries which utilize the staple product as an input. In addition, the growing population sustained by employment in the leading sector creates *final demand linkages*, such as for housing and consumer goods (compare Fig 4.3). The degree to which the gradual diversification or 'maturing' of the domestic economy actually takes place, and the geographical pattern in which it does so, depends quite significantly on the technology associated with the staple's exploitation. In Canada, the wheat staple, which called for investment in an extensive transportation system, was particularly critical in promoting an integrated but diversified domestic economy (Wolfe 1968).

The danger of resource-based development is that a nation, or some large region of it, may get stuck in a *staple trap*. Although initial growth of the staple industry may be associated with booming markets, the long-run prosperity of staple export regions is bound to decline unless economic diversification is successful. The resource may become exhausted, or be displaced in world markets by cheaper sources of supply or synthetic substitutes. Its markets may simply stagnate as consumption patterns change in favour of more highly manufactured products. The very success of the staple industry may encourage such narrow specialization of capital investment and narrow vision in local institutions (firms, unions, public agencies) that attitudes and the structure of the economy become inflexible and all the more vulnerable to changes in the external environment (see p. 155). Such problems are far more common in peripheral nations or regions than in the more economically diversified core. Canada's position in the contemporary global economy is somewhat ambiguous, as is Australia's, for they exhibit many of the characteristics of industrialized core nations yet retain some of the vulnerability of peripheral staple producers (McCann 1987).

Vance (1970) and Burghardt (1971) have analyzed the spatial and institutional framework within which transportation and settlement systems evolved in support of the long-distance trading economy of North America. Following exploratory voyages by European (core) 'merchant adventurers', colonization took place in regional patterns which reflected the nature of the terrain and the staple exploited (contrast the isolated coastal fishing communities of Newfoundland with the incipient urban system of early coastal Pennsylvania). Trading links with overseas markets gradually clustered in a few coastal centres (Vance's 'points of attachment', Burghardt's 'gateway cities') where the larger groups of core-based merchants established their base of operations in the periphery (Fig. 3.5). Their control of the staple export trade (and the return flow of European manufactures) was furthered by the development of transportation routes penetrating the hinterland. Continued population growth and the eventual achievement of political independence by the New World nations provided the conditions whereby these eastern gateway cities were able to transform their functions and market orientation and emerge as *nodal centres* (see p.177), within the emergent core regions of the USA and Canada. The processes of gateway city formation and the articulation of the hinterland economy through them were repeated as the agricultural frontier entered the Great Plains (see p.91 and Fig. 3.5).

The sociostructural framework of frontier expansion and resource development involved the fusion of atomistic individual decision-making with a matrix of order imposed by economic and political interests in core regions (initially European and later domestic). Railways

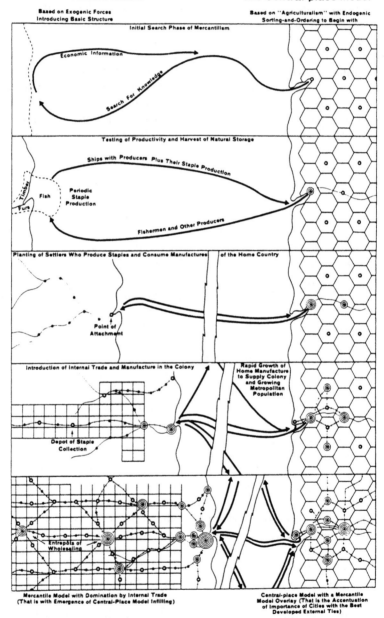

The Mercantile Model

Based on Exogenic Forces
Introducing Basic Structure

The Central-place Model

Based on "Agriculturalism" with Endogenic
Sorting-and-Ordering to Begin with

Initial Search Phase of Mercantilism

Economic Information

Search For Knowledge

Testing of Productivity and Harvest of Natural Storage

Ships with Producers Plus Their Staple Production

Timber
Fish
Furs

Periodic
Staple
Production

Fishermen and Other Producers

Planting of Settlers Who Produce Staples and Consume Manufactures | of the Home Country

Point of
Attachment

Introduction of Internal Trade and Manufacture in the Colony

Rapid Growth of
Home Manufacture
to Supply Colony
and Growing
Metropolitan
Population

Depot of Staple
Collection

Entrepôts of
Wholesaling

Mercantile Model with Domination by Internal Trade
(That is with Emergence of Central-Place Model Infilling)

Central-place Model with a Mercantile
Model Overlay (That is the Accentuation
of Importance of Cities with the Best
Developed External Ties)

Figure 3.5 The evolution of a settlement system based upon long-distance trade in staple products.

Source: Vance, J.E. 1970. *The Merchant's World: The Geography of Wholesaling*. Englewood Cliffs, NJ: Prentice-Hall, Figure 18.

made continental interiors accessible on a large scale in the 19th century, and railway companies often had the most immediate economic interest in attracting immigrants and stimulating staple production to create traffic for their expanding networks. Meinig (1962) compared the experience of two similar grain-producing regions, the transportation systems of which were organized within differing political frameworks. Railway construction in the Columbia Basin of the northwestern USA involved intense private-sector competition; in the region north of Adelaide (South Australia), it proceeded under centralized government control. Several differences in the route geometry of the two railway networks and the nature of the service which they offered to farmers were evident. But there were also notable similarities, reflecting the railways' common goal of promoting a prosperous staple economy and the common lack of influence of the farming communities. In both cases, 'most of the ultimate decisions were made by groups outside the region whose interests were by no means coincident with those to whom services were to be granted' (Meinig 1962, p.412).

In contrast to the economic diversification which staple production has stimulated in colonies of settlement (if not necessarily within every region of them), its impact in *colonies of exploitation* has invariably resulted in a staple trap. These tropical territories were ones in which the production of a mineral or agricultural staple took place on the basis of a repressive social order imposed by Europeans on an existing population, or on an imported one (of slaves or indentured labourers) where native people were few in number (see Ch. 4). Mines or plantations formed *enclaves*, islands of capitalist production and its associated technological and economic infrastructure in a region whose indigenous economic development was, at best, severely disrupted (at worst, destroyed). Taaffe, Morrill & Gould (1963) have conceptualized the sequence of transportation-system and urban-centre development typically associated with this form of European penetration (Fig. 3.6). (The development of staple enclave economies prior to the railway era frequently took place on islands, as in the Caribbean, or in coastal regions.)

Auty (1985) analyses the basis for the differential outcome of export staple production in colonies of exploitation as opposed to colonies of settlement. He attributes the poor performance of tropical staples to 'a combination of more rigid production functions, lower staple substitutability, weaker spatial integration and elitist social structures' (Auty 1985, p.20). The larger fixed capital investment of plantations compared to frontier family farms or early lumbering operations reinforced dependence on a single crop, even in the face of weak demand. Very low wages, maintained if necessary by importing new workers, and the restricted civil freedoms of labourers resulted

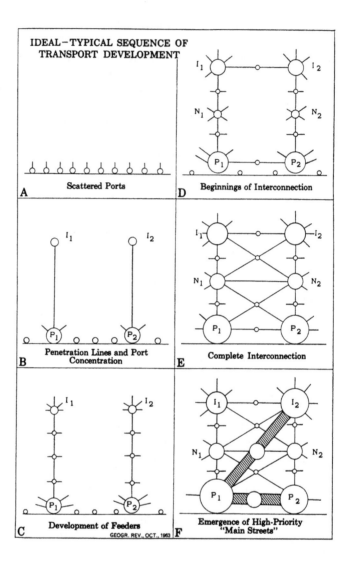

Figure 3.6 Idealized sequence of transportation system development associated with European penetration of tropical colonies.
Source: Taaffe, E.J., R.L. Morrill, & P.R. Gould 1963. Transport expansion in underdeveloped countries. *Geographical Review* 53; Figure 1.

in feeble final demand linkages and little incentive to develop productivity-enhancing backward linkages. The limited spatial extent of the enclave economy and its limited demand for a regionally integrated transportation system further reduced its potential for diversification.

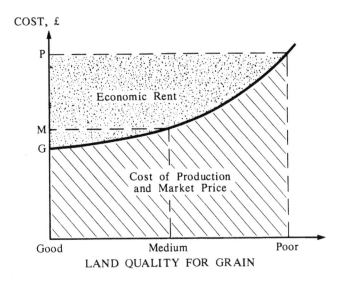

Figure 3.7   The basis of economic rent: Ricardo's model.

The geography of economic rent
Within a geographically extensive economy, regional differentials in the nature and quality of natural resources become increasingly significant. Theories of international trade focus primarily on the benefits of exchanging dissimilar specialized commodities. The related concept of *economic rent* can be used to analyse the impact of qualitative differentials between alternative sources of a given commodity. These arise out of the basic material characteristics and spatial dimensions of the Earth, and so economic rents are universal, although the form they take depends on the political economy of the society in question.

Neoclassical models of rent formation derive from similar but independent analyses by Ricardo (1817) and von Thünen (1826). Both writers argued that within any given market for natural resources, prices reflect the revenue necessary to provide the marginal supplier with a 'normal' profit. Other (intramarginal) producers thus receive a net income which provides more than normal profits, the basis for which is the superior fertility of their land (Ricardo) or their more favourable location with respect to the market (von Thünen). Economic rent so defined corresponds to a form of 'differential rent' within Marxist theory (Harvey 1982).

Ricardo's interest was aroused by the impact of population growth on the early development of the Industrial Revolution in England. He saw that the expanding demand for grain was bringing increasingly infertile land into agricultural production. Within a capitalist economy of private land ownership, the principal beneficiaries of rising grain prices (reflecting higher production costs at the margin) were those landlords who reaped windfall gains as their economic rent increased (Fig. 3.7). The principal losers were the capitalist manufacturers rendered uncompetitive by the rising level of subsistence wages they were forced to pay their employees (who, in real terms, remained as poor as before). The immediate policy conclusion which Ricardo drew from his analysis was that Britain's protectionist Corn Laws should be repealed (which they were in 1846; see Ch. 4) to allow grain imports to lower prices, reduce rents and defuse a prospective Malthusian crisis.

Von Thünen's analysis of the rents derived from differential location stemmed from his experience as a farmer confronted by rudimentary farm-to-market transportation. His theoretical model of *The isolated state* (1826) incorporated assumptions of an *isotropic* plain (one of uniform environmental characteristics), a single market, one mode of transportation, and profit-maximizing farmers. For a given crop, with geographically uniform production costs, farmers' net incomes will vary in direct relation to the cost of transportation, and hence to their proximity to the market (Fig. 3.8). The margin is thus defined spatially, and in this context economic rent is a *location rent*.

Chisholm (1979) assembles copious evidence of the rôle of location rent in determining the distribution of agricultural activity at a variety of geographical scales, from the village to the world-economy. The Enga farmstead (Fig. 2.4) illustrates that the spatial logic of von Thünen's model is present in non-market societies, in which the minimization of effort in crop production replaces revenue maximization as the basis of land-use patterns. Peet (1969) demonstrates the systematic spatial structure of Britain's food imports in the 19th century. Muller (1973) reveals that geographical variations in net agricultural revenue per hectare in the USA in the mid-1960s were, above all, related to distance from the largest regional market, megalopolis of the North East (Fig. 3.9).

Analysis of economic rent can readily be extended from the agri-cultural context to other renewable and to non-renewable resources, although the timing of rental income needs more attention. Calculation of rents involves such considerations as the optimum revenue stream to be derived from a multiyear crop (as in forestry), the optimum depletion rate of a stock resource (such as an oilfield), and the exploration costs involved in finding mineral deposits (when only one prospect in a hundred may result in commercial extraction). Moreover, in comparison to agricultural resources many other natural resources are geographically localized, and hence more susceptible to monopolization by a single owner or cartel. The power of OPEC in the 1970s reflected the fact that the largest and most prolific oilfields in the world were controlled

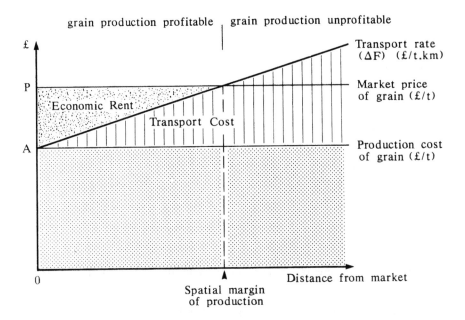

Figure 3.8 The basis of economic rent: von Thünen's model.
Optimum land use within a multi-crop context can be decided for any given location by calculating which crop yields the highest economic rent at that distance from the market, using the formula:

$$R = E (P-A-FK) \qquad \text{in which}$$

R = economic rent per unit of land (£/ha); E = yield per unit of land (t/ha); P = market price per unit of output (£/t); A = unit production cost (£/t); F = transport rate (£/t.km); K = distance (km).

by a unified political group which had considerable scope to set prices. High oil prices yielded OPEC members both economic rent, based on global differentials in the cost of production, and an element of *monopoly rent*, derived from their concentrated ownership of least-cost oil. Less successful cartels, in commodities such as uranium, have existed at various times in recent decades. Although OPEC had lost its market power by the late 1980s, the potential exists for its re-assertion as world oil reserves are further depleted.

There is no immediate reason to expect the geographical distribution of differential rents based on resource productivity to resemble that based on market accessibility. The former reflects the global pattern of climatic, soil, and vegetation conditions, and the random distribution (from a human perspective) of stock resources; the latter reflects patterns

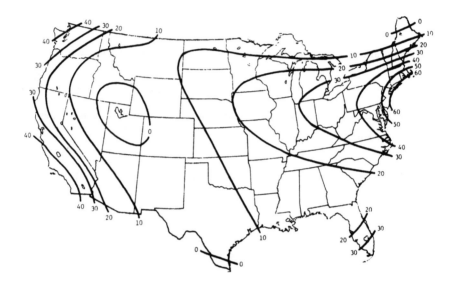

Figure 3.9 Net income ($) per acre for farms in the USA, 1964: quartic trend surface.
Source: Muller, P.O. 1973. Trend surfaces of American agricultural patterns: a macro-Thünian analysis. *Economic Geography* 49; Figure 8.

of social interaction which have changed through time. Yet the two are not as dissociated as might appear. Much depends on looking for suitable evidence at an appropriate geographical scale. Although Ricardo (1817, p.33) wrote of 'the original and indestructible powers of the soil,' any land with a long history of agricultural production becomes a human product to some extent. Depending on how the soil is treated, fertility can be increased or depleted. Insofar as its improvement involves effort, such as the spreading of manure, land that is close by will benefit more than land at a distance. This can be corroborated at the scale of the farmstead (Fig. 2.4) and the rural town (Chisholm 1979), which until the 19th century in Europe was, and in many developing countries still is, a major source of night-soil. Intensive agriculture around major markets can be seen as a response, therefore, both to high location rents and high fertility (compare Fig. 8.8). At a national scale, core regions are frequently those whose relatively favourable environment for agriculture has, over time, given rise to higher standards of living and levels of capital formation than elsewhere. Certainly, natural advantages have been reinforced by political decisions, but in this context one notes the rôle of London and the South-East within England, the Paris Basin within France, southern Ontario within Canada, and Victoria and New South Wales within Australia.

The geographical relationship between resource quality and market access is less clear-cut for non-agricultural resources, but spatial patterns are discernable. At the global scale, the strong link between stock resource consumption and industrialization meant that markets developed first in north-west Europe and northeastern USA. The limited scale of early technology meant that many small, local deposits (especially of coal and iron ore) could be utilized. As the volume and commodity composition of industrial demand expanded, more distant supplies were sought, as necessary, to augment or replace original sources. Invariably, these were of higher grade to offset higher transportation costs. The escalating raw-material demands of the capitalist industrialized nations after 1945 triggered a worldwide search for new resource deposits, resulting in massive investment in production capacity and major advances in bulk transportation technology (Beaver 1967). Increasingly, therefore, the global market for non-renewable resources has become dominated by low-cost producers (whose economic rents have in many cases been diluted by excess capacity in their industry) rather than by those close to major markets (many of which are approaching exhaustion).

Resources in an international industrialized economy

The population/ resource ratio

The early 1970s were marked by growing fears of an imminent and widespread crisis of food and resource availability. The mid-1980s were characterized by concerns about the difficulties facing resource producers as a result of low prices and excess production capacity. This dramatic reversal of global prospects was partly a reflection of changing ideological perspectives (Blaikie 1986), but major shifts in some of the underlying material conditions justified a less pessimistic outlook. In a period of rapid and radical change it is difficult to predict how long resource supplies will remain adequate. For instance, although global food production has more than kept pace with the growth of population in recent decades, there are concerns about the sustainability of the agricultural output which has been achieved, and the prospects for supporting further increases in per capita consumption. On the other hand, the significant productivity increases achieved even since 1960 in mechanized Western agriculture, and the prospect of yet greater efficiencies associated with the commercialization of biotechnology, suggest that the globe has substantial reserve capacity to feed projected population increases. Similarly, the threat of serious shortages of non-renewable resources has currently almost disappeared, for a combination of reasons which are reviewed below. Yet doubts remain about the long-term preparedness of industrialized societies to adapt to scarcer and more costly oil supplies. And concern has dramatically increased about the cumulative environmental consequences of the prevailing technologies which underlie economic growth. If the *global* adequacy of resources is not an immediate cause for concern, it is quite clear that the populations of certain regions continue to face very severe threats to their subsistence and capacity for resource-based economic growth.

The world's population passed the 5 billion mark in the mid-1980s, having approximately doubled since 1950. Its growth and growth rates are unequally distributed across the continents, so that concern about the sustainability of any given total has to be addressed on a regionalized basis (Fig. 3.10). The industrialized economies of the USSR, Europe (East and West), and North America are increasing their populations very slowly (some countries even face population decline in coming decades). Over the century 1950–2050, their combined share of the world's total will shrink from 29.2 to possibly 7.2%. In contrast, Africa's share could rise from 8.7 to 23% over the same period, and South Asia's from 28.4 to 38.2%. The implications of these dynamics for the development prospects of Third world nations are discussed more fully in Chapter 10.

Renewable resources   Global increases in food production have become fundamentally dependent on increased yields, as the scope for expansion of the agricultural frontier has shrunk. Increased fertilizer usage and improved strains (especially of the Green Revolution crops: wheat, maize, and rice) have played a major part in raising output and have been particularly significant in assuring basic food supplies in the world's two most populous countries, China and India. The productivity increases in industrialized capitalist countries have been magnified by heavy government subsidies to agriculture which have disproportionally benefitted large, well capitalized farmers. Yet increased economic pressure to pursue mechanized monoculture has given rise to adverse environmental impacts, notably soil erosion and groundwater depletion. In the Third World, deforestation is perhaps the most severe problem. Its causes vary regionally, but more intensive land use within tropical forests leading to shorter fallows, increasing pressure on fuelwood supplies, and vigorous forest clearance to accommodate commercial agriculture (a feature of the Amazon basin in particular) all play a rôle. Even in a developed country such as Canada, poor past

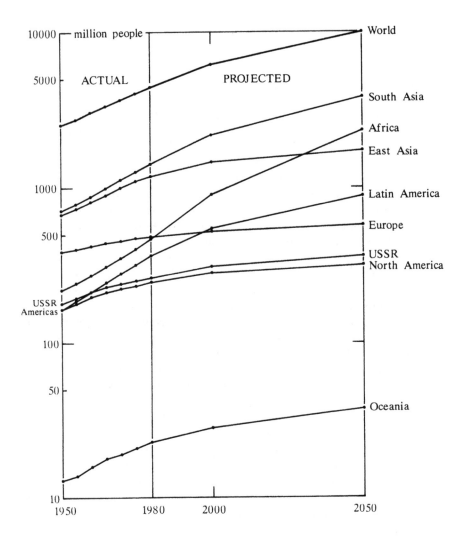

Figure 3.10 Global population growth, 1950–2050.
Source: derived from Woods, R. 1968. Malthus, Marx and population crises. In *A World in Crisis: Geographical Perspectives*, R.J. Johnston & P.J. Taylor (eds) Oxford: Blackwell; Table 6.1.

practices of forest management have resulted in impending regional shortages of commercial timber.

Non-renewable resources Paradoxically, the potential scarcity of stock resources is currently less of a threat than that of flow resources. This is partly because human sustenance is less directly dependent on them (although recent famines have indicated the vital rôle of mechanized transportation in supplying food deficit areas), and partly because the production systems which transform them are rarely dependent on a single (exhaustible) source. The dramatic improvement in the projected availability of non-renewable resources between the early 1970s and the mid-1980s can be explained in a number of ways (*Financial Times* 29 August 1986):

(a) The predictions of imminent resource depletion encouraged a new wave of exploration and investment which resulted in many new, large projects coming into production. Most of these were in the Third World (except in the oil industry, where politically stable regions such as the North Sea were prominent).

(b) The slowing of economic growth in the capitalist world in the wake of rising oil prices undermined the pessimistic extrapolations of resource depletion rates which featured in such studies as *The limits to growth*. It has become increasingly evident, moreover, that economic growth in advanced societies is less resource-intensive than it used to be. This reflects the growing share of services in total output (see Ch. 8), increasing sophistication in the design of structures and equipment (which allows one to accomplish more with less), the nature of new technologies (notably microelectronics) with their strong emphasis on miniaturization, and the much greater emphasis on conservation and recycling, which has been fostered particularly by rising energy prices. (Note that these trends have been uniformly weaker within the industrialized state-socialist nations.)

(c) The surplus production capacity which has resulted from the sequence of events described above has been exacerbated by a variety of institutional policies. Third World governments have maintained or expanded their output, even in the face of falling prices, to meet their foreign exchange needs and service their debts. TNCs and other buyers have frequently encouraged resource developments in new supply regions or nations as a means of diversifying their sources (and of increasing their bargaining power in the process).

Whereas the foregoing considerations suggest that the economic growth of resource users may not be seriously hindered by material shortages in the foreseeable future, there are causes for concern. The corollary of plentiful and cheap non-renewable resources is that those nations whose income is heavily dependent upon selling them (and these are primarily in the global periphery, as we have noted) face severely constrained development prospects. Moreover, the degree to which transportation technologies remain wedded to petroleum suggests that oil shortages (or disruptively steep rises in the price of petroleum products, which is the economic equivalent) will reappear sometime within the next 40 years. When this happens, poorer nations and social groups will be the earliest and most severely affected.

The vulnerability of production systems

Increasing economic and technological specialization and increased social dependence on the infrastructures associated with them carry risks which contemporary societies are apt to overlook. One explanation for this is the undoubtedly greater resilience to unexpected 'shocks' which advanced industrial societies possess in comparison to less developed ones. The ability to summon resources from afar, thanks to modern transportation and communications technologies, diffuses the *local* severity of disruptive events within a broader economic system, although some of the burden of adjustment may be felt by vulnerable groups elsewhere (Warrick 1983). At a global scale, the manner in which financial institutions reacted to the OPEC price increases and recycled petrodollars to support a changed pattern of capital investment, while by no means uroblematic, demonstrated a much greater capacity to cope with shocks to the international economy than was evident in the 1930s (see p.102). Yet the technocratic (over-)confidence of 'modern' societies and the weakness of ecological perspectives within economic analysis (Marxist or liberal) creates an atmosphere in which risks and hazards tend to be discounted, not least those arising out of or affecting environmental systems. The mind-set that habitually thinks of 'fixing things' views major disruptive events as unexpected discontinuities, discrete in space and time, rather than as symptoms of explicable processes which call for an appropriate and strategic social response (Hewitt 1983a).

The dependence of industrialized agriculture on large inputs of fossil fuels and growing evidence of the adverse ecological consequences of this form of production constitute a major area of concern. Brown (1985) notes the reduced annual growth in world grain ouput which followed the OPEC oil-price increases, as farmers were forced to economize on fertilizers and mechanized equipment. Wessel (1983) cites research which indicates that two-thirds of US cropland is experiencing a net loss of soil, and that erosion is heavily concentrated in the most productive areas where corn and soybeans (both row crops) predominate. The Science Council of Canada (1986) provides evidence of comparable losses in soil productivity in that country. The high levels of fertilizer consumption with which farmers in Western Europe and North America have sought to boost output in the face of soil degradation have created severe problems of water contamination in many regions. A British report issued in 1986 estimated that within 20 years, capital spending of £200 million and annual outlays of £10 million will be needed to keep nitrate concentrations within acceptable limits (*Financial Times*

16 December 1986). Denmark became the first country to legislate a reduction of fertilizer applications for this reason (*Financial Times* 19 November 1986).

Awareness of the ecological (and the economic) costs of prevailing agricultural practices is stimulating change, such as the growing adoption of zero-tillage (which reduces soil erosion and conserves moisture) in the grainlands of North America. Reduced agricultural subsidies in the EEC and the USA will also lessen the environmental stresses associated with high-energy-input farming (see p. 180). The productivity increases which more environmentally compatible biotechnologies are expected to yield in coming years may well be sufficient to match global population growth; but given the institutional channels through which they will be diffused (Buttel *et al.* 1985), their benefits may not be available to many cultivators in many less developed countries.

The rôle of nuclear power generation in industrialized economies also raises acute questions of the vulnerability of contemporary production systems. The initial promise of civilian nuclear programmes was that they would reduce the eventual crisis of fossil-fuel depletion by providing electricity that was cheaper and less environmentally polluting than alternative (essentially coal-based) technologies. The dangers associated with the nuclear fuel cycle, both of catastrophic events such as the Chernobyl explosion and of low-level radiation; the technical and political problems of long-term waste disposal (not least the question of where to put it); and the economic costs of decommissioning old reactors have all tended to be addressed with a technocratic optimism that acceptable solutions will be found as the need arises. There is no denying the basic dilemma which societies face over nuclear energy, nor that the attractiveness of different options varies considerably between nations with different resource endowments and different political economies (contrast the USA with France, for instance). East European authorities argue that nuclear energy is much less damaging to the environment than burning brown coal would be. The state of California, on the other hand, has ruled that no further nuclear plants will be built in its jurisdiction until the long-term waste disposal issue has been adequately resolved. Whichever stance a nation takes towards nuclear energy, the global issues of economic and environmental vulnerability associated with its use remain.

The implications of global climatic change for the smooth functioning of economic systems are reviewed in Chapter 12. There are other natural events, however (some of them undoubtedly precipitated or magnified by human action), the impacts of which deserve more serious attention than they generally receive. The vulnerability of the San Francisco area, and particularly Silicon Valley, to a major earthquake within the next 30 years is a good example. It also demonstrates the significance of social variables in determining the outcome of naturally hazardous events. The computer technologies underlying San Franciso's rôle as a major financial centre and Silicon Valley's research and development (R&D) activity are highly vulnerable to a major seismic shock. The predicted earthquake is therefore liable to generate a business crisis with international repercussions, over and above the massive damage to regional economic infrastructure and disruption of civil life that it will entail. Yet precautionary planning against such an event is frequently hampered by a culture in which anything which threatens to reduce present property values may prompt litigation. Japan, another seismically sensitive economy, spends over one hundred times more per head annually than California on earthquake preparation (*Toronto Globe and Mail* 30 November 1985). The Ecuadorian earthquake of early 1987, which ruptured the country's oil export pipeline and led to an immediate suspension of international debt repayments for the rest of the year, and the Armenian earthquake, which cost the USSR fully half of its national economic growth in 1988, confirm the vulnerability of the global economy to events in the material environment on which it is founded.

Urbanization

Behind many of the examples of environmental pressures and the vulnerability of production systems reviewed above lies the fact of the increasing size of the world's urban population. This is essentially a Third World development in that the industrialized economies of East and West have already attained high levels of urbanization, have achieved low rates of population increase and, in the West, have even begun to see a deconcentration of population away from their largest cities (see Ch. 8). In contrast, urban growth in most Third World societies is being intensively fuelled by large absolute increases in total population and continuing rural to urban migration. It is indicated in Figure 3.11 that, by the year 2000, 80% of the world's largest cities (over 10 million inhabitants) will be in developing countries, and that many of these cities will be larger than any city except New York and Tokyo has ever been.

The physical and organizational infrastructures required to meet the material needs of large metropolitan areas evolved haphazardly in most European and North American cities in the 19th century (see Ch. 4). The necessity of adequate public investment in sewage and water supply systems was finally accepted on health grounds, but the provision of food and energy and the transportation systems which deliver them were rarely given explicit public attention. Fortunately, most large cities grew slowly by contemporary Third World standards, they did not become as large and, above all, their public and private institutions were in a better financial position to make the necessary capital investments.

Urban agglomerations with more than 10 million inhabitants: 1950, 1975, and 2000

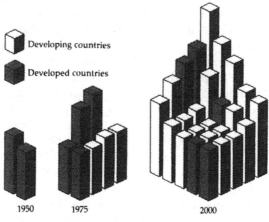

Developing countries

Developed countries

1950   1975   2000

1950	(millions)		(millions)
New York, northeast New Jersey	12.2	London	10.4

1975			
New York, northeast New Jersey	19.8	London	10.4
		Tokyo, Yokohama	17.7
Mexico City	11.9	Shanghai	11.6
Los Angeles, Long Beach	10.8	Sao Paulo	10.7

2000			
Mexico City	31.0	Sao Paulo	25.8
Tokyo, Yokohama	24.2	New York, northeast New Jersey	22.8
Shanghai	22.7		
Beijing	19.9	Rio de Janeiro	19.0
Greater Bombay	17.1	Calcutta	16.7
Jakarta	16.6	Seoul	14.2
Los Angeles, Long Beach	14.2	Cairo, Giza, Imbaba	13.1
		Manila	12.3
Madras	12.9	Bangkok, Thonburi	11.9
Greater Buenos Aires	12.1	Delhi	11.7
Karachi	11.8	Paris	11.3
Bogota	11.7	Istanbul	11.2
Tehran	11.3	Osaka, Kobe	11.1
Baghdad	11.1		

Figure 3.11 Urban agglomerations with more than 10 million inhabitants: 1950, 1975, and 2000.
Source: World Bank 1984. *World development report 1984*, figure 4.3.

The prospects of comparable infrastructure provision to cope with contemporary urban growth in developing countries are not bright, and this raises questions about the environmental sustainability, not to mention political and economic viability of, for example, a Mexico City of 31 million.

Douglas (1983, p.103) outlines the problems confronting Mexico City's water supply. Situated in an internal drainage basin at an altitude of 2260 m, the metropolitan area has long had to rely on groundwater. Accelerating abstraction since 1939 has led to subsidence of up to 20 cm per year and cumulative subsidence of 7 m in places. This 'has greatly increased the flood hazard . . . [and] required massive urban drainage works.' As the city continues to grow, both stormwater disposal and the provision of additional drinking water from distant river catchments involve crossing the basin's watershed. By the 1990s, new supplies could be coming from a location 200 km away and 200 m lower, and pumping this water would require the energy output

> . . . of six 1,000-megawatt power plants, . . . construction of [which] would cost at least $6 billion, roughly half the annual interest payments on Mexico's external debt . . .. The city is thus faced with three rising cost curves in water procurement – increasing distance of water transport, increasing height of water lift, and rising energy costs (Brown & Jacobson 1987, p. 52).

Saudi Arabia has been able to afford a seawater desalinization plant and a 180 km pipeline to suppy Riyadh (710 m above sea level) with enough water to handle a population of 1 million (*Financial Times* 28 October 1985), but few Third World states can match its per capita income!

Conclusion

Human beings have physical needs, and civilization requires a material culture. However much the driving force of the capitalist world-economy appears to have become financial movements, 'uncoupled' from resource production (see Ch. 1), and however much some aspects of core societies appear to have become 'post-industrial' (see Ch. 8), the interactions between people and their natural environment have a fundamental bearing on the course of economic development and the range of economic opportunities available in particular places. This chapter has traced the changing nature of society–environment relations as the technological capability to harness resources and conquer space has increased. By the late 20th century, the evidence that human actions in pursuit of economic growth are triggering major changes to the global environment can no longer be ignored (see Ch. 12). First, however, it is necessary to understand the roots of today's challenges. The specific historical and geographical configuration of the incorporation of the Earth's resources into the capitalist world-economy, and the evolution of the social structures through which this was accomplished, are the subject of the next chapter.

4 The evolution of the modern world-economy

Time, scale, and structure

Any writer reviewing the evolution of the capitalist world-economy is forced 'to play the honest broker between two contradictory yet essentially correct views' (Supple 1972, p. 35). One methodological perspective looks for patterns in the process of development from which theoretical explanations of functional relationships may be constructed. Another sees a sequence of specific events which can only be understood in the context of particular institutional settings. In practice, all meaningful historical explanation has a conceptual or theoretical framework, even if only implicitly. Similarly, the value of a theory, however logically sound, ultimately rests on its ability to account for past or emerging characteristics of the real world. The world we live in inherits a history which could theoretically have been very different: to apply theoretical insights to help shape its future, we need to appreciate the long-term implications of its actual past.

We follow Wallerstein's (1976) argument that, once one moves beyond the essentially self-sufficient local society, the only satisfactory scale for starting to analyse economic activity is that of a world-system (see Ch. 1). No denigration of the past achievements of Chinese or any other non-European civilization is implied by acknowledging that the modern world is, in very many significant ways, a product of developments which originated in Europe. Hence we focus first on some of the critical economic and social developments which gradually transformed the geography of Europe during the 600 years prior to the Industrial Revolution. Then we analyse the causes and consequences of the European involvement with other parts of the world, which gradually expanded from the 15th century. A geographical appraisal of the Industrial Revolution in Great Britain and North America, and of

Table 4.1 Exchange types and distribution systems.

Distribution system	Level of commerce	Exchange type	Spatial characteristics	Distribution of the élites	Where found
extended network system	uncommer-cialized	dyadic, direct	Fig. 4.1a	not relevant (undifferen-tiated)	independent 'tribal' societies
bounded network system	uncommer-cialized	polyadic, direct	Fig. 4.1b	concen-trated, rural (settled or mobile)	'Chiefdoms', 'feudal' manors
solar central-place system	partially commer-cialized	administered market	Fig. 4.1c	concen-trated, urban (settled or mobile)	pre-modern colonies, developing states and empires
dendritic central-place system	partially commer-cialized	monopolistic market	Fig. 4.1d	absent from peasant region (settled)	'Peripheries' of 'modern' economic systems
interlocking central-place system	fully commer-cialized	competitive market	Fig. 4.1e,f	even, rural and urban (settled or mobile)	'Cores' of 'modern' economic systems

Source: adapted from Smith, C. A. 1976. Exchange systems and the spatial distribution of elites: the organization of stratification of agrarian societies. In *Regional analysis; Vol. II: Social systems* C. A. Smith, (ed). New York: Academic Press; Tables 1 and 2.

the radical social transformation (notably massive urbanization) which it produced, follows. The migration of millions of Europeans to settle in other parts of the globe, and the rapid expansion of Euro-centred colonial empires in the 19th century, completes the background for examining the basic structure of the capitalist world-economy at its zenith in the early decades of the 20th century.

Note throughout this chapter the interplay of environmental conditions, social and political structures, and the way in which men and women interpreted the world and their position within it, as formative influences on economic development.

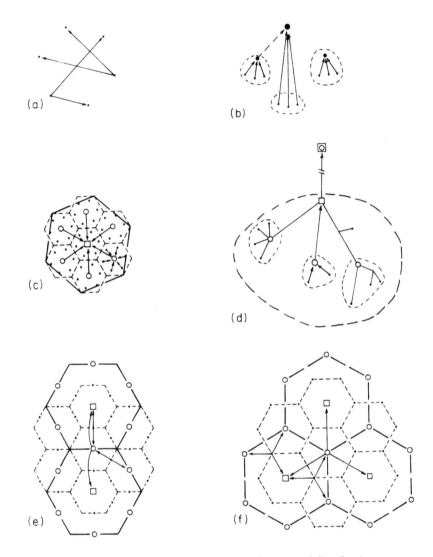

Figure 4.1 The spatial expression of differing exchange and distribution
systems.
Six ideal systems of nodes or central places:
a An unbounded network system
b A bounded, hierarchical network system
c A solar central place system
d A dendritic central place system
e,f Interlocking central place systems
Dashed lines show boundaries of the system; arrows show
relationship of lower-level to higher-level centre(s); size of nodes or
central places shows order in the hierarchy, open symbols are
market centres, closed symbols places without markets.

Source: Smith, C. A. 1976. Exchange systems and the spatial distribution of elites: the
organization of stratification of agrarian societies. *In Regional analysis; Vol. II: Social systems*,
Smith C. A. (ed). New York: Academic Press; Figure 1.

The world of pre-industrial Europe

The growth of towns and market functions

Many scholars in different disciplines have attempted to give a convincing explanation for the emergence of urban settlements. Nevertheless:

> It is doubtful if a single autonomous, causative factor will ever be identified in the nexus of social, economic and political transformations which resulted in the emergence of urban forms ... [W]hatever structural changes in social organization were induced by commerce, warfare or technology, they needed to be validated by some instrument of authority if they were to achieve institutionalized permanence. (Wheatley 1971, p.318).

Regularities in the spatial structure of urban systems need to be explained within this broad context (Fig. 4.1, Table 4.1). Moreover, one cannot understand urban centres independently of the wider society in which they are set: whether a feudal town or a modern metropolis, they are best seen as 'fields of action integral to some larger world and within which the interactions and contradictions of that larger world are displayed with special clarity' (Abrams 1978, p.3).

The Roman world–empire disintegrated more thoroughly after the 3rd century AD in its northern European periphery than its Mediterranean core, where much of the capital investment of the Roman period was retained in usable form. As a result, the economic revival of southern urban centres, especially those accessible to the channels of maritime trade, came well in advance of similar renewal further north. Luxury items from Asia and the Middle East reached Europe through the hands of Italian merchants based in autonomous city-states, such as Venice, Genoa, Florence, and Milan. The market for these goods consisted of the élite groups who dominated the social structure of Western Europe through their command over its land and predominantly rural population. By the 13th century, the rich merchants who dominated this lucrative trade added another source of demand.

Whether their wealth derived from agrarian surpluses or from the monopoly profits of long-distance trade, the élite groups satisfied their requirements through urban markets. Exotic items from afar were only a fraction of their consumption compared to crafted goods and buildings of local origin and, above all, food and other essential products of the rural economy. Craft workers, the building trades, and personal servants of all varieties constituted a large proportion of the slowly growing urban population which, together with those for whom they worked, provided a gradually increasing market for the agricultural produce of the surrounding countryside. Until the onset of the Industrial Revolution, trade and marketing was the mainstay of the urban economy, whereas manufacturing, from at least the 15th century onwards, exhibited a growing orientation to rural resources.

The economy of pre-industrial Europe was *partially commercialized* (Table 4.1) in that the vast majority of the population lived on the land and provided for its own subsistence. The limited volume of traded output entered distribution systems which were subject to a considerable degree of political regulation. Markets had two basic functions; to supply the urban population with food and other necessities from the surrounding rural economy, and to link the local economy to the wider world. In both cases the purchasing power of the élite groups was the primary stimulus to market transactions. By controlling the mechanics of marketing through institutional regulation of prices and the frequency of markets, these groups were able to concentrate wealth derived from the surplus production of the rural economy in their own hands. They were assisted in this by the physical constraints facing any peasant who might have thought of seeking alternative markets. Only gradually, as the demands of larger urban centres required it and transportation improvements made it possible, did the insular pattern of local, regulated food markets give way to freer and more widespread trade linking cities with regions of specialized agricultural production.

The marketing of goods which linked local with national or international networks of supply naturally exhibited different characteristics. Merchants held the key rôle as middle men between buyers and sellers who lacked direct intercommunication. Frequently able to act at the local level as a monopolist supplier, their prices ensured ample profits, even when the undeniable risks of long-distance trade were accounted for. Peasants were regular buyers of staples such as salt and, less frequently, of necessities such as cooking utensils and iron goods. The wealthy strata of urban society also purchased such items, but in addition created the demand for a wide variety of luxury goods. Markets for these commodities were infrequent: transactions often took place at an annual regional fair, to which itinerant merchants brought their wares.

Relative location and regional economic specialization
The emergence of nodal mercantile cities, seen first in Italy, next gathered pace in Flanders, the Rhinelands, and southern Germany. Together, these regions constituted the prosperous core areas of Europe in the 14th century. But by the 16th century, important changes in the economic geography of Europe and its expanding world-system had come about. As the spatial extent of trade increased, shifts took place in the location of strategic centres. Slowly rising population levels and standards of living in northern Europe gradually increased the volume and variety of economic activity there. In the Baltic, merchants based

in the cities of the Hanseatic League approached the prosperity of the Italian merchants trading in the Mediterranean. When Spain suddenly acquired a central position in the European economy on the basis of its bullion inflow from the New World, the Seville–Flanders–Baltic axis combined with the Seville – Genoa – eastern Mediterranean axis to create an 'external', maritime channel of intra-European trade. This severely reduced the economic significance of the traditional overland routes between northern and southern Europe. The accessibility of coastal areas in Britain and Western Europe to the new trans-oceanic trade reinforced this shift in Europe's economic centre of gravity. Note, however, that the strategic position of Flanders was enhanced by these new orientations, in which it was geographically, and became functionally, central. Antwerp emerged as the commercial focus of the expanding world-system.

The gradual emergence of significant concentrations of urban popu-lation in a still predominantly agricultural world, and the consolidation by the 17th century of a few economically advanced core states within a rapidly expanding arena of trade relations, both highlighted at different scales the influence of distance and market accessibility on economic activity. The demands of a city such as London (with a population of 200000 in 1600) for fresh milk and livestock generated patterns of agricultural specialization, with perishable products concentrated close to the urban market and more readily transportable supplies, notably grain, further afield (Wrigley 1978). Transportation costs simultaneously affected the regional intensity of production for market concentrations at a continental scale. Notably, Polish landowners began to specialize in grain to supply the urban, mercantile economy of the Netherlands, whose rural areas concentrated on more intensive mixed farming.

Geographical specialization in agricultural production prior to the Industrial Revolution was parallelled by trends in the production of other commodities. Trade in manufactured goods, organized by town-based merchants on the basis of production dispersed throughout their rural hinterlands, became increasingly characterized by regional specializations reflecting localized resource endowment. The wool textile industry, the largest employer, became concentrated in the cottages of agricultural villagers at least partly to escape the constraints of guild regulations governing urban craftworkers. Other important activities, such as mining, metal-working, and salt production were predominantly rural because their raw materials and energy supplies were located there (Nef 1964). By the 16th century, in England especially, the volume of wood used for fuel in industrial processes from metal-working to brewing contributed to a serious depletion of timber supplies. The resulting inflation of timber prices greatly increased the attraction of coal as a source of heat, and this tended to increase the concentration of 'heavy' industry in selected mining regions. Meanwhile, growing volumes of trade within and between the nations of Europe stimulated growth in related industries.

The developing economic rôle of the state

The growing prosperity of the Netherlands, England and, to a lesser degree , France, in the 16th century reflected more than their favourable location on European trade routes. In contrast to Spain, whose Hapsburg monarch sought not only to strengthen national cohesion and economic power but also to maintain dynastic control over a diverse and scattered empire, these three nations saw a marked and relatively undistracted strengthening of the power of the *nation-state*. As a result of changes in the domestic political balance of power, interests of state became increasingly defined in terms of economic advancement rather than dynastic or religious objectives. One of the important advantages which a nation-state enjoyed over an empire was its scale:

> The political centralization of an empire was at one and the same time its strength and its weakness. Its strength lay in the fact that it guaranteed economic flows from the periphery to the center by force (tribute and taxation) and by monopolistic advantages in trade. Its weakness lay in the fact that the bureaucracy made necessary by the political structure tended to absorb too much of the profit, especially as repression and exploitation bred revolt which increased military expenditures (Wallerstein 1976, p.15).

In contrast, the smaller size and greater cultural homogeneity of a nation-state made it easier to create a strong administration, able to exert pressure both on local political units internally and on other states externally. The process of national integration, or the consolidation of the power of central authorities, had a number of geographical consequences, although specific states responded in differing ways to similar challenges. Capital cities such as London and Paris increased considerably in size and significance as markets for goods and services. The need to finance a growing civil service and a permanent military establishment meant that governments increasingly saw the nation as a corporate business enterprise. North & Thomas (1973) argue that the age of *mercantilism* was born of the urgent need for revenue to cope with recurrent fiscal crises, as government expenses grew much faster than traditional sources of revenue could accommodate.

The essential philosophy of mercantilism was quite simple, and not inappropriate in its original setting. National prosperity depended on a positive balance of trade, which built up the national wealth as measured by its holdings of precious metals. These ensured the maintenance of a sound domestic currency which could support a gradual expansion of economic activity. Among the specific measures which governments took to generate a positive trade balance were *navigation laws*, requiring goods to be carried in domestically owned, rather than foreign, ships. The export of raw materials for use by foreign manufacturers was prohibited and tariffs were levied on imported goods. As European nations began to carve out colonies overseas, their mercantilist policies attempted to ensure that the wealth of their colonial economies contributed exclusively to economic growth 'at home' (Kammen 1970).

The expansion of Europe overseas

Powerful economic interests fuelled the spectacular expansion of the Europe-centred world-system in the century and a half following 1500, but other motives were active also. Moreover, there were continuities between developments within Europe and those further afield. The example of Spain illustrates these features very clearly (Meinig 1969). The nationalistic and religious crusade which expelled the Islamic Moors from Spain was continued almost immediately by the *conquistadores* in South America. Later, the political and religious ramifications of the Reformation in northwestern Europe played an important rôle in promoting European settlement in North America, notably in the actions of the Pilgrim Fathers. But it is as misleading to mask the economic motivations behind European expansion as it is to suggest that all other factors were merely subservient to them.

The incentives to expand trade, and the monopoly profits which could be derived from it, were felt strongly in the emerging nation–states of Europe and in the core region (Spain) of the Hapsburg empire. The initial objective of European merchants who pioneered new sea routes was to reach the traditional source of exotic products in South-East Asia. In the 16th and early 17th centuries, Portuguese, Spanish and Dutch traders initiated regular commerce, exchanging Oriental spices, silks, and cottons for European gold and silver. With the European 'discovery' of the Americas at the close of the 15th century, a second major trans-oceanic commerce developed, one which developed in a much more complex fashion for a variety of reasons.

The highly profitable Oriental trade involved commodities with a high value and low unit weight being carried halfway around the world. The Americas did not immediately offer anything so attractive, but their greater accessibility to Europe broadened the options of trading groups seeking a profit. Less densely populated, and at a lower level of technological development than the Orient, intertropical America was a region where military force could extract desired resources more cheaply than peaceful trade. The Spanish conquest and colonization of central Mexico and the central Andes was directed at those regions where the pre-existing civilizations 'offered not only the labour that the Spanish wanted in large quantities to exploit the areas within which they settled, but also wealth in the form of precious metals . . . and worked land.' Native knowledge of mining and agricultural techniques enabled 'the Spanish to build up their level of economic activities in their newly won colonies very quickly' (Odell & Preston 1973, p.121). As Frank (1975) argues, it was the regions whose gold, silver, and tropical agricultural products made them appear the richest parts of the New World to contemporary European eyes which suffered the greatest economic and social debilitation. In contrast, the temperate regions of northeastern North America and the grasslands bordering the River Plate, which initially appeared to offer little of value to Europe, were comparatively neglected until opened up under very different social and economic conditions as *colonies of settlement* rather than as *colonies of exploitation* (see Ch. 3).

Although there were significant cultural and political differences between the colonies of Catholic Spain and Protestant Great Britain, the influence of the resource base on the type of colonial economy which developed in a region is evident in the contrast between British America's northern and southern colonies. While these two regions were alike in presenting Europeans with an abundance of land and a relative scarcity of labour, the two colonial environments differed climatically. Sugar (with its molasses and rum by-products), tobacco, and later cotton, could be grown in the south, but the profits of the commercial planters depended on a captive and low-cost labour supply. *Plantation agriculture*, in which a single commercial crop was grown in a large-scale, hierarchically organized production unit, dependent upon slave labour, became the dominant response to these environmental and economic parameters. Between 1500 and 1870, plantations in the southern USA, the Caribbean, and Brazil 'consumed' 10 million African slaves. The slave population of Britain's US colonies was the only group to grow by natural increase; elsewhere the death rate was so high that the labour force could be maintained only by constant imports (Grigg 1974). Soil exhaustion was also a typical result of plantation monoculture.

British America's northern colonies promised little prospect of great wealth, and were settled predominantly by people who owned and worked their own land, albeit within the context of a market economy. Given the limited range of commercially attractive resources, economic development was triggered by the rôle which colonial traders carved out for themselves in the commerce linking Europe and the intertropical colonies of exploitation. The north developed as a supply base for capital goods (notably ships) involved in trans-Atlantic trade, and for a range of staple products (food, timber, basic manufactures) required by the narrowly specialized extractive economies to the south.

The economic growth and diversification of British America's northern colonies confirmed the growing evidence in Europe: states which gained the most from colonial empires were not those which grabbed the richest prizes but those with the strongest and most adaptable domestic economies. During the 17th century, the dynamics of intra-European trade (and the fringe benefits of privateering!) effectively reduced Spain to no more than a funnel through which New World bullion was channelled in exchange for goods produced elsewhere in Europe (Sella 1974). Even before the Industrial Revolution, Great Britain was becoming the core of commercial activity in the widening world-economy.

The Industrial Revolution in Great Britain

Background

Towards the end of the 18th century the pace of economic development in Great Britain accelerated, perceptibly and permanently. Within 50 years, such fundamental changes had taken place in resource use, population growth and its geographical distribution, and the volume and variety of industrial production, and the term 'revolution' is certainly justified. Yet to understand the Industrial Revolution in Britain, it is critical to recognize that these changes were profoundly *evolutionary*. Developments in different aspects of British life, involving the cultural and political as well as the economic sphere, began to interact in such a way that a cycle of self-reinforcing *circular and cumulative causation* (Myrdal 1957) was established (see Fig. 4.3). It is misleading to try to *rank* contributory factors or to claim any logical *necessity* for what in fact took place (Mokyr 1985). Our objective here is to review a number of the major 'ingredients' of Britain's Industrial Revolution and thereby stress that economic development is a complex process, not amenable to 'quick fixes'; a necessary perspective in assessing the industrialization attempts of contemporary Third World nations (see Ch. 11).

The resource base Wilkinson (1973) argues convincingly that a growing shortage of land and renewable resources constituted a major challenge to economic activity in 18th century Britain. Many of the innovations which made up the Industrial Revolution came in response to price changes reflecting the relative environmental scarcity of resources which could be substituted for one another. The steady expansion of coal mining and the related expansion of brick production were responses to the growing shortage of wood as a fuel and a structural material. Increased pressure on the supply of agricultural land raised its price, the rents derived from it, and the cost of its products. In a pre-industrial economy still dependent on land for its food and raw materials, the ramifications of these developments were far-reaching. Higher food prices, especially from the 1750s, gave landowners the financial incentive to expand and improve their production. Higher agricultural rents provided much capital for investment in other sectors. The higher cost of land-intensive goods hastened the substitution of horse transportation (2 ha of land under hay were needed to feed each horse) by canals and, later, railways. The substitution of woollen clothing by cotton represented an exchange of British land under sheep for American and Indian land under cotton. By the end of the century, a growing shortage of water-power sites hastened the adaptation of the steam engine to power factories.

Population and agriculture   By the early 18th century, Britain was feeling the benefit of both escape routes from the 'Malthusian trap', increasing productivity of resource use and migration. At a time when the domestic population stood at around 6 million, there were already over half a million people of British descent living abroad (North & Thomas 1973). The quality and quantity of farm production was slowly rising, especially in southeastern England, which enjoyed the best climatic conditions and proximity to Holland. Dutch immigrants introduced the innovation of replacing fallow with legumes and root crops, which both improved soil fertility and provided winter fodder on a previously unattainable scale.

The incentive and freedom for large landowners to adopt new crop rotations depended on their having control of their land. The 18th century saw a rapid acceleration of the *enclosure movement*, by which the agrarian economy of common-field agriculture was dismantled in favour of one of privately owned farm units. This process, in which many villagers lost their customary right to use common land, has been characterized as 'a plain enough case of class robbery, played according to fair rules of property and law laid down by a parliament of property-owners and lawyers' (Thompson 1968, pp. 237–8). That enclosure brought wealth to a few, and greater insecurity if not always greater poverty to many, is indisputable. But it is equally clear that the prosperity of the large landowners derived overwhelmingly from the greater productivity of agricultural activity on enclosed holdings, and very little from the partial expropriation of small landowners and residual common fields (McCloskey 1975b). A better-nourished population began to grow after the middle of the century, and although food prices rose, so did opportunities for employment in a growing economy. It is significant that Britain's agricultural output was increasing as it embarked upon its Industrial Revolution. Domestic food production largely kept pace with the onset of sustained population growth (although food imports certainly increased). Rather than suffering the misery of agricultural involution, the bulk of the British population enjoyed a slowly rising standard of living which widened the market for manufactured goods.

The state and the economy   In contrast to most of continental Europe, by the beginning of the 18th century Britain possessed a unified national market, relatively free from government regulation. Internal barriers to trade, in the form of tolls, or different local currencies and legal systems, had been eliminated. Many of the medieval guild regulations which restricted the freedom of action of economic agents, governing such things as apprenticeships, wage levels, and labour mobility, were simply ignored. At the same time, mercantilist policies were applied externally to the benefit of British traders and manufacturers (Flinn 1966).

The contribution of the colonies   Britain's overseas trade was growing faster than the national economy as a whole throughout the 18th century. In particular, the various 'triangular' trades of the North Atlantic, involving, for instance, the export of British manufactures, the shipment of slaves from Africa to the plantations of the 'Greater Caribbean' region, and the import of plantation products (sugar, rum, tobacco, etc.) to Europe, were a source of capital formation and a school of business experience. The captive colonial markets were critical for a few British industries, notably cotton (see p.92), but the profits of colonial trade overall were dwarfed by gains realised from productivity improvements and rental income in the domestic economy (O'Brien 1982).

Transportation   Markets could not be expanded without improved transportation, and in this sphere too the critical developments took place *within* the British economy. A variety of undertakings increased the capacity and the productivity of roads and inland waterways well in advance of the Industrial Revolution itself. The problem of maintaining road surfaces under increasingly heavy traffic had to be tackled institutionally, through the creation of turnpike trusts, before there was adequate incentive to tackle it technically (Wilkinson 1973). Canal construction, actively promoted by landowners and major manufacturers, made many regions newly accessible and greatly improved the reliability of water transportation by guaranteeing a given draught. (The water level of many rivers was notoriously unpredictable, the more so as water-power sites multiplied). As always, reduced transportation costs helped producers to reach wider markets and pursue the benefits of greater specialization.

Culture and institutions These economic developments took place within the matrix of a specific culture, in the absence of which many of them would either not have emerged or not have had their dynamic effects. For example, the creation of national markets and the diffusion of innovations in agriculture or industry would have been harder to achieve in a less literate society. The existence of an English-language Bible, and Protestant encouragement to read it, was significant in this respect. The churches played a more explicit rôle in providing general education to children of the lower classes, preparing them, often all too ideologically, for 'various useful and practical arts' in the economy of industrial capitalism (Flinn 1967, p.16; quoting a Sunday school annual report). In comparing Britain to other parts of Europe, it is often difficult to pinpoint *why* different attitudes prevailed. Why were the landed gentry of England so much more inclined to invest in and experiment with agricultural innovations than their counterparts in France? Why were they more inclined to trust their capital to an emergent banking system? We lack definite answers to such questions, but this should make us all the more sensitive to the cultural and institutional context of economic development.

Summary The Industrial Revolution in Britain was preceded over a period of many years by a gradual evolution of attitudes and institutions, and a steady rise in the volume and variety of economic activity. Some key manufacturing sectors, notably the iron industry, had been faced with impending resource crises (in this case, the diminishing availability of wood for charcoal production) which had been overcome by technological developments. But a more general rise in demand, as real incomes slowly increased and cheaper interregional transportation widened markets at home and abroad, tended more and more to stimulate innovation in the production process of a wide range of goods. The development of these innovations involved a revolution in the social organization of industry at least as profound as in its technology of resource transformation (see p.84).

Geographical changes in the British economy

Population distribution Prior to the Industrial Revolution the population of Britain, although more urbanized than that of any other state except Holland, was predominantly rural, and the majority of the labour force was engaged in agriculture. Around 1750, one quarter of the population lived in towns of at least 2500 inhabitants. London, with 675 000, alone contained almost 10% of the national total. The next largest city, Bristol, a base for much of the trade with the New World, trailed far behind with 50 000 (Daunton 1978). Regional variations in the density of the

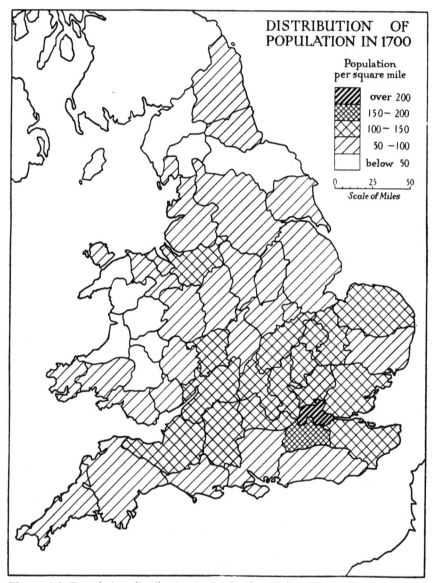

Figure 4.2 Population distribution in England & Wales, 1700 and 1801.
Source: East, W.G. 1936. England in the Eighteenth century. In *An historical geography of England before A.D. 1800*, H.C. Darby (ed) Cambridge: Cambridge University Press; Figures 83 and 84.

agrarian population matched variations in the physical resource base. The most prosperous agricultural regions were the lowlands of the south and east, which were also the areas best served by navigable rivers, along which cheap transportation of bulk commodities facilitated the

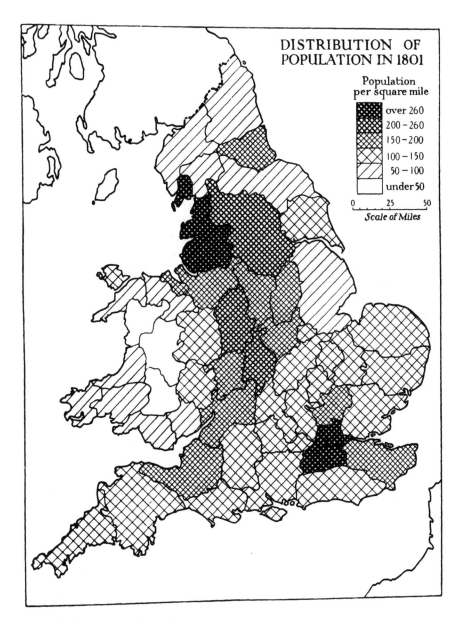

DISTRIBUTION OF
POPULATION IN 1801

Population
per square mile

over 260
200 – 260
150 – 200
100 – 150
50 – 100
under 50

0   25   50
Scale of Miles

spread of specialized production to sustain the huge market of London. Conversely, even before the Industrial Revolution, rural domestic industry was tending to concentrate in the northern and western regions where agriculture offered relatively poorer prospects. Rapid expansion of factory production in the North and the Midlands after 1780 brought about substantial changes in the distribution of population (Fig. 4.2). The proportion of the labour force engaged in agriculture had dropped to 35% by 1801, although the absolute numbers continued to rise as the national population continued its unprecedented growth.

Regional economic specialization   From the middle of the 18th century, a national network of canals made many upland regions accessible by water transport for the first time, and it was these areas which were best endowed with the coal (of growing importance as an energy source, especially for the iron industry) or undeveloped water-power sites required by the growing number of factories. The woollen industry, long the largest source of British manufactured exports, increasingly concentrated in western Yorkshire as production was progressively mechanized. The cotton industry, a much more recent branch of production, in which innovations could be introduced with greater ease, rapidly became concentrated in south-east Lancashire, in the hinterland of the Liverpool cotton merchants. These various regional specializations were well developed *before* a national network of railways began to take shape in the late 1830s, which helps to explain why British industry prior to the First World War was so heavily concentrated in a group of coalfield regions (see Ch. 8).

The North American sequel

As the Industrial Revolution was getting under way in Britain, the original coastal colonies of British North America fought for and won their independence. Among all the factors which contributed to colonial dissatisfaction with British rule and the ineptitude of the British government in handling colonial grievances, economic issues were certainly prominent. Britain's mercantilist policies had discouraged the growth of manufacturing in North America. Release from these constraints did not bring about immediate changes in the economy of the USA, but gradual developments prepared the way for a period of spectacular and fundamental reorientation of the economy between approximately 1830 and 1870 (the second Kondratieff upswing). This took the northeastern states well along the road to matching Britain's industrial achievements.

The colonial commercial economy
From the beginning, the economy of British North America had been commercial, and the culture urban, even though only 5% of the US population lived in towns of 2500 or more inhabitants in 1790. A handful of port cities (Boston, New York, Philadelphia, and Baltimore) acted as

THE NORTH AMERICAN SEQUEL

the nodal centres, or 'gateway cities' (Fig. 3.5), through which British manufactured goods were channelled into the back country, and to which the surplus produce of the rural settlers was despatched for sale. The trading activities which formed the basis of the colonial economy came increasingly under the control of US merchants in the years after 1776. Profits tended to be re-invested in commerce in the northern states, or invested in plantations in the South; little went into manufacturing, where British imports still dominated. Labour tended to be in short supply in the cities, given the attractions of an open frontier not too far to the west.

The industrial transformation
By the time of the Civil War, the northern states had developed a substantial industrial economy (and the southern states a structurally similar but much smaller one). Many ingredients of the British experience were evident in North America, but events tended to move much faster in the New World. This was especially obvious with respect to population growth, boosted by mass immigration after the 1830s. Domestic demand for manufactured goods mushroomed with it. Transportation improvements integrated this growing market and promoted regional specialization. The opening of the Erie Canal in 1825 provided the first major commercial link to the interior beyond the Appalachians, and it firmly secured New York's position as the pre-eminent port gateway to the USA. Very soon thereafter, railways were built inland from the port cities to secure their control over an expanding hinterland. A broad regional division of the antebellum economy evolved, in which the plantations of the South concentrated on producing cotton, primarily for export to Britain; the North East developed an increasingly diversified manufacturing sector based on its coal and water-power resources; and the farming frontier of the 'Old North West' (the Ohio River basin) concentrated on producing foodstuffs (Bruchey 1976). The trading complex of the northern port cities increasingly reoriented its focus from Europe towards the domestic market, which was expanding rapidly in size and geographical scope.

The process of change
The transformation of the mercantile city, the economy of which was dominated by its trading complex (and in which manufacturing activity was primarily limited to *servicing* that complex, with ships, barrels, paper, etc.), into a manufacturing centre, the trading (wholesaling) establishments of which increasingly focused on *distributing* the products of its factories, was an instance of circular and cumulative causation (Pred 1966, 1973). Gradual expansion in 'the wholesale-trading complex' generated demands for a range of related goods and services from

other sectors: this is the *multiplier effect* (Fig. 4.3). Demand for the output of these related sectors gradually increased so that either (a) it became profitable to manufacture locally goods which were previously imported, or (b) existing producers were able to develop greater functional specialization and economies of scale. Either way, regional demand achieved new thresholds which justified the diversification of the local manufacturing sector.

Innovation

This second process involved the implementation of new production techniques or the creation of new products that enabled manufacturers to expand their markets by offering cheaper or better quality goods (often both). Pred interprets the evidence to suggest that the environment which best fostered innovation was found in the rapidly growing port cities of the North East. Here, information about technological developments in Britain was most readily available (in press reports or the accounts of visitors); an expanding market justified frequent investment in new, 'state of the art' machinery; and an expanding

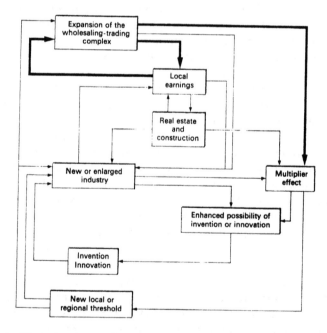

Figure 4.3 The circular and cumulative feedback process of local urban size-growth for a single large US mercantile city, 1790–1840. Heavy lines indicate the most important relationships.

Source: Pred, A. 1977. *City-Systems in Advanced Economies: Past Growth, Present Processes and Future development Options*. London: Hutchinson; Figure 2.8.

manufacturing sector provided a growing concentration of people looking for solutions to technical problems, many of which could be applied in different branches of production. As manufacturing processes became increasingly specialized and interdependent, 'technological disequilibrium' between sequential stages of production focused innovative attention on critical bottlenecks.

Technology and the North American resource base
The Industrial Revolution in North America took place within a very different resource context from that of Great Britain. Manufacturing grew against the background of a national economy which was undergoing constant geographical expansion, and this only reinforced the basic contrast between the North American and the European population/resource ratios. Borchert (1967) has imaginatively portrayed the link between the progress of settlement and the evolution of metropolitan centres across the USA, and epochs in the technology of transport and industrial energy by which resources were distributed and transformed (Fig.4.4). By 1870, manufacturing cities had developed throughout what became known as 'the Manufacturing Belt' (Fig. 4.5). From their initial rôle (at various previous dates) as commercial nodes,

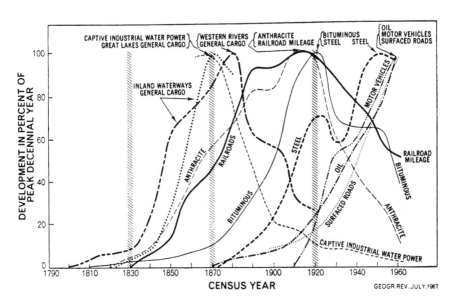

Figure 4.4 Indicators of the growth and decline of transport and energy technologies in the USA, 1790–1960. Note that peak values tend to concentrate around 1870 and 1920.
Source: Borchert, J.R. 1967. American metropolitan evolution. *Geographical Review* **57**; Figure 1.

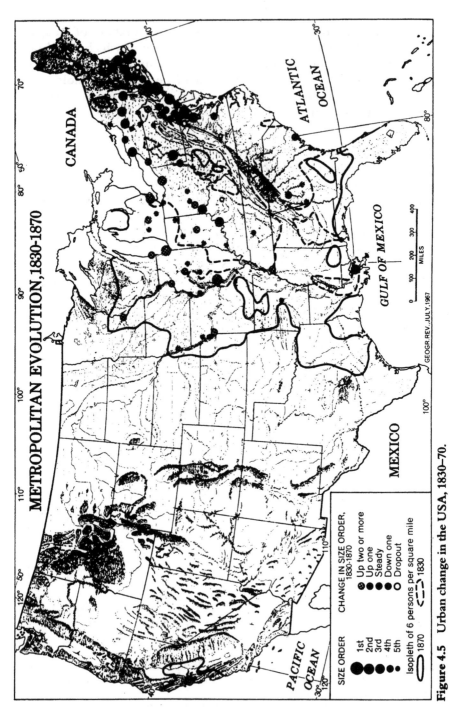

Figure 4.5 Urban change in the USA, 1830–70.

Source: Borchert, J. R. 1967. American metropolitan evolution, *Geographial Review 57*; Figure 8.

and then as regional manufacturing centres, these cities became the bases from which established firms went on to develop nationwide markets as the railway network spread west and south.

It has been argued that the technological progress of US industry, which was increasingly commented on by 19th century European visitors, was the result of the relative scarcity of labour in that country. Both in the cities, beyond which the open frontier beckoned, and on the frontier itself, there would seem to have been a considerable incentive to invest in labour-saving machinery. It is not at all clear, however, that innovation in US industry prior to 1870 was any more rapid or labour-saving than in Britain. A much more critical factor appears to have been the British interest in conserving relatively more expensive raw materials. David (1975) cites studies of the cotton spinning and the woodworking industries which indicate that US equipment was more capital-intensive, but it also required (or wasted) considerably more raw material than the equipment used in British industry.

The cultural and political context
The social and economic divergence between the northeastern and the southern states has already been noted. The negative consequences of the plantation economy became increasingly apparent in the decades before the Civil War. The South was a more conservative society, dominated by landed interests and rigidly segregated on racial lines. Not only was there less interest in the development of manufacturing, but the inferior social and economic status of slaves meant that a growing population did not translate into mass markets for industrial products. In contrast, the North was a more open society, receptive to new ideas and less rigidly stratified. Further north still, in central Canada, the dominant socio-economic group tended to maintain its traditional mercantile interests, and many early manufacturing initiatives were undertaken by entrepreneurial individuals from the USA (Tulchinsky 1977).

The Industrial Revolution: its institutional and urban context

The factory and the firm
Most of Britain's rapidly emerging industrial towns began life within an institutional framework that was ill adapted to their needs. The basis of local government was invariably still rural parishes, and the urban population had almost no say in electing the national government. Not until the 1830s did the sentiments and geography of British politics begin to reflect the emergence of the urban industrial economy, and to address the issues which it raised (Briggs 1968). Citizens of the USA were

generally able to observe and reflect on the course of industrialization in Britain before they were caught up in it themselves. Leaders of the newly independent nation debated the merits of encouraging manufacturing (Grampp 1970). Against arguments that it would create employment and give agricultural producers a stable domestic market, Jefferson warned of the social consequences in Europe, where he attributed the absence of genuine democracy to the emergence of mass urban populations of poor factory workers.

The factory system offered both technical and managerial advantages to the entrepreneurs who established it. The invention of machines to carry out specific manufacturing processes greatly increased the output of the individual worker. To achieve economies of scale in energy consumption required concentrating machines (and hence workers) at a single power source (initially a water wheel). The gathering of workers under one roof overcame the growing diseconomies of supervision and transport costs associated with traditional domestic manufacturing, in which a merchant distributed raw materials to households and collected the finished goods for market (Flinn 1978). The factory system developed first and most rapidly in the British cotton textile industry, precisely because this was a relatively new branch of manufacturing (having been discouraged by legislation protecting the woollen industry) and hence one in which radically new ways of organizing production could be introduced with less resistance from established interests.

The concentration of machines into a factory profoundly changed the human relations of production. Household industry was part of the agrarian economy; workers would be impoverished but not immediately faced with starvation if unemployed, and the entrepreneur's control of production was flexible and incomplete. But under the factory system, the capitalist faced substantial fixed costs in plant and equipment, making him much more 'a prisoner of his investment' than before. Workers faced an even more traumatic transformation, wrenched from the rhythms and security (diminishing, admittedly) of the agrarian economy and reduced to the status of a 'machine hand':

> Now the work had to be done in a factory, at a pace set by tireless, inanimate equipment, as part of a large team that had to begin, pause, and stop in unison – all under the close eye of overseers . . . The factory was a new kind of prison; the clock a new kind of jailer (Landes 1969, p.43).

Conflict in the city

In the industrial cities of 19th century Britain and the northeastern USA the explosive growth of population around a nucleus of factories produced a highly charged social environment. The differential prosperity of the capitalist entrepreneur and the impoverished worker was dramatic: never before had labourers been so massed in a geographically restricted area. The prevailing ideology disassociated the private goals and decisions of individual capitalist entrepreneurs from the collective aspirations and interests of the broader urban society, in which wage-labourers were numerically dominant. Their resistance to the combined onslaught of mechanization and proletarianization took a variety of forms, but it was hindered by external opposition and internal fragmentation. Until the 1830s, British trade unions existed in a 'disputed area on the border of legality' (Thompson 1968, p.552), so that even the threat of legal action against them was inhibiting. Early union activity among craftsmen in US cities was directed as much against the immigrants who threatened job security as against employers. Unionism was always more fragile in the USA than in Britain and much more vulnerable to disintegration in periods of economic depression (Mohl & Betten 1970). Nevertheless, by the late 19th century, working-class resistance in the largest industrial cities of the USA was effective enough to contribute to the decentralization of manufacturing which began to take place (Gordon 1978).

Working and living conditions in the early industrial city were uniformly grim (Vance 1977). Population growth depended on constant immigration to make good deaths by disease. Some early factory owners had accepted responsibility for housing their employees, but it soon became the norm for workers to be left to the mercy of speculative builders. Densely packed dwellings with few or no services, but yielding a healthy rental income, were built without any municipal control. Eventually it was the threat to public health and fear of rioting by the impoverished masses which prompted British authorities to legislate improvements in the living conditions of the urban poor. Conditions were similar in comparable cities in the USA. New York was described in 1832 as 'one huge pigsty' by a former mayor (quoted by Mohl & Betten 1970, p.93).

The geography of resource exploitation and the great migrations

Population–resource dynamics in Europe

After a long history of very gradual expansion, marked by considerable fluctuations, the population of Europe totalled 130 million in 1750; but by 1800 it had reached 185 million, then 266 million by 1850, and 401 million by 1900. Already the most densely populated continent, Europe increased its share of the world's inhabitants from one-fifth to one-quarter over this period, while simultaneously losing a further

28 million who settled other continents (Armengaud 1976). This fundamental shift in the dynamics of population growth, the *demographic transition* (Fig. 3.1), has attracted various explanations: one emphasizes improvements in nutrition made possible by more reliable and better quality food supply; another invokes the impact of better sanitation and improved medical knowledge, although their contribution to reducing mortality came later in urban areas. Explanations of the subsequent drop in the birthrate are less clear-cut. Rising confidence that children would survive to adulthood, increasing female employment outside the home, and attachment to rising standards of living which would be threatened by excessive family size, are three frequently cited factors.

The interrelated developments in industry, agriculture, transportation, and population growth which constituted the Industrial Revolution in Britain began to appear in continental Europe in the 19th century. Some geographers have explained this as a diffusion process, in which 'waves of modernization' originating in Britain gradually 'washed' across Europe, transforming its society and economy in their wake. At any given date during the period up to 1914 one could certainly observe an approximate north west–south east gradient in levels of economic development, but the Industrial Revolution accentuated an existing disparity – it did not create it. Britain, France, Holland, and Sweden were nation–states which had long pursued, within their varied capacities, mercantilist policies to strengthen their national economies. In contrast, modern Germany and Italy were still politically fragmented. The geography of industrialization in Europe most resembled a scatter of 'islands' of modernity (often associated initially with a domestic armaments industry, in which 'most countries felt the need to keep up to date' (Saul 1972, p.54), and later with numerous railway workshops) punctuating a generalized decline in the level of technical competence away from the English Channel.

Continental Europeans found it easier to catch up with the advances of British industry than those of British agriculture. Despite many similarities between the agrarian societies of England and the continent during the feudal period, profound differences in its legal and political systems had differentiated English society at a very early date. English real-property law allowed much easier transfer of land between individuals and at the same time treated inherited property as indivisible, meaning that a family holding was usually taken over by the oldest son. Continental practice, in contrast, retained a stronger communal conception of landownership, which inhibited change of proprietorship or use, and was characterized by *partible* inheritance, meaning that a family holding was divided among members of the next generation (Macfarlane 1979). Whereas the enclosure movement had completed the institutional and economic basis for 'progressive' agriculture in England, advances in agricultural productivity on the continent were hindered by the excessive fragmentation of land holdings and the weight of levies in cash or kind demanded by the landowning nobility. Britain's growing population was fed by an increasingly productive agriculture, which as early as 1840 employed only one quarter of the labour force. On the continent, population growth tended rather to intensify the poverty of a majority still tied to the land (Fig. 4.6).

One indication of the large-scale expansion of the 'poverty culture' of the era was the rapid expansion of potato cultivation. The displacement of bread grains by the potato in the family diet involved not merely a nutritional retreat but also 'a loss of standing, a loss to be delayed as long as possible' (Lis & Soly 1979, p.180; quoting Morineau). The geographical and institutional channels by which the plant was initially diffused throughout central Europe have been traced by Denecke (1976). First confined to the botanical gardens of 'scientists' (1560–1650), then spreading to private gardens as a regular food plant (1650–1740), the potato was vigorously promoted as a field crop after 1740, not least by governments who saw its high yield per hectare as an answer to the increasingly marginal subsistence of a rapidly increasing population (Fig. 4.7). As poor regions became increasingly dependent on potato monoculture for their staple diet, they became increasingly vulnerable to potato blight. Nowhere was the link between impending starvation in Europe and migration overseas more explicit than in Ireland, where the

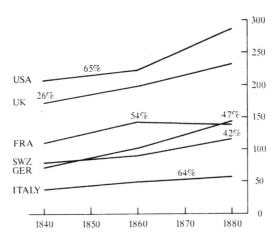

Figure 4.6 Index of the level of agricultural development in Western Europe and the USA in the 19th century. Percentages record the proportion of the working population in agriculture at the time of the earliest available census.
Source: compiled from Bairoch, P. 1973. Agriculture and the Industrial Revolution 1700–1914. In *The Fontana economic history of Europe, vol 3: The Industrial Revolution 1700–1914*, C.M. Cipolla (ed). Glasgow: Fontana; Tables 1 and 2.

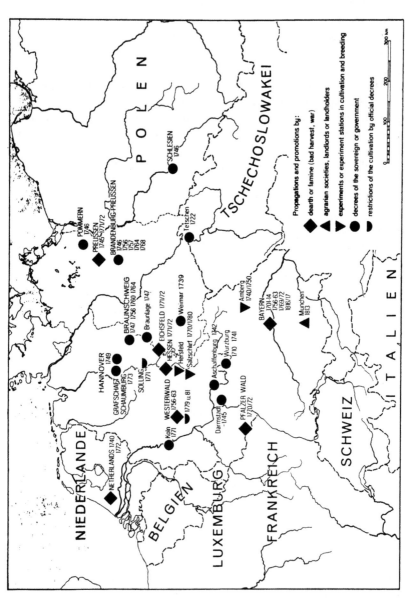

Figure 4.7 Propagation and promotion of the cultivation of the potato as a field crop in central Europe in the 18th century.

Source: Denecke, D. 1976. Innovation and diffusion of the potato in central Europe in the seventeenth and eighteenth centuries. In *Fields, farms and settlement in Europe*, R. H. Buchanan, R. A. Butlin & D. McCourt (eds.). Belfast: Ulster Folk and Transport Museum; Figure 5.

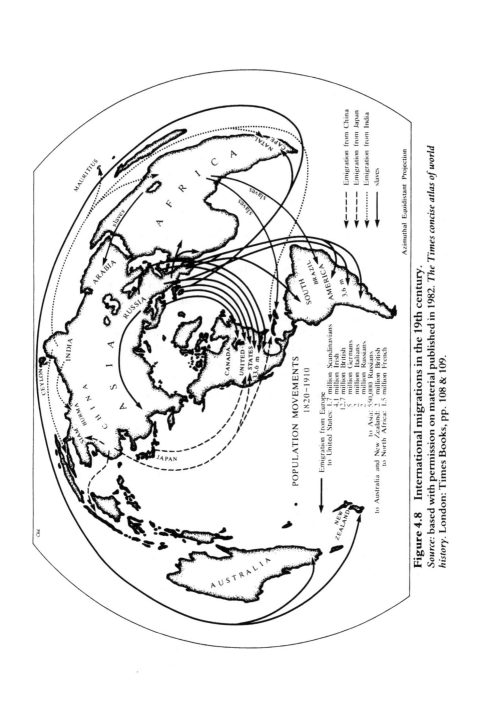

Figure 4.8 International migrations in the 19th century.
Source: based with permission on material published in 1982. *The Times concise atlas of world history.* London: Times Books, pp. 108 & 109.

population was reduced by a quarter in only six years (from 8.5 million in 1845 to 6.5 million in 1851) as a result of the potato famine.

Frontiers become nations

Approximately 2 million Europeans (mainly Britons) emigrated to overseas colonies in the 18th century. As many again (mainly from central and southern Europe) left in a single year (1910) at the peak of the mass exodus that followed, from 1840 to 1914. European traders had traced the geographical framework of the capitalist world-economy: the settlement of tens of millions of Europeans across the temperate regions of the remaining continents consolidated the cultural and economic supremacy of that system (Fig. 4.8). The causes of this unparalleled uprooting of humanity were complex, but the *push factors* of adverse conditions in the migrants' homelands were generally much more decisive than the *pull factors* of the opportunities (often dimly perceived) in the territories to which they moved. Yet the two sets of factors were linked. As the European population multiplied and began a sustained growth of per capita resource consumption, it both increased the pressure on the resource base of its own continent (pushing many to leave) and created a demand for the marketable resources of other parts of the world (attracting many to exploit these opportunities). Improvements in transportation on land and at sea brought about by the steam engine were a vital ingredient of this process. So too was the export of capital to develop the infrastructure of the newly settled territories.

Wood was generally the scarcest staple commodity in pre-industrial Europe, but it was bulky and expensive to transport in relation to its value. Having gone far into northern Europe to obtain timber by the close of the 18th century, the British Navy and British industrialists were forced (by Napoleon's blockade of the Baltic) to look to their North American colonies (eastern Canada) for alternative supplies. The viability of a lumbering industry so far from the market was ensured by a massive tariff on Baltic supplies, which protected the high-priced North American product from competition in peacetime. A lumbering industry which 'mined the forest' was always on the move and not a large-scale source of permanent settlement. Wheat, however, was a staple with very different characteristics. Growing markets in the urban concentrations of Europe and repeal of the British Corn Laws in 1846 stimulated steadily rising imports from newly settled land in North America, made increasingly accessible by the expanding railway network and the Erie and St Lawrence canals. By the late 19th century, the volume of European demand and the declining cost of ocean transportation were such that settlers on the Argentine pampas and in South Australia could also sell grain to the Old World. The introduction of refrigeration and other technological developments widened the scope for agricultural exports from these regions (Youngson 1965).

By 1870, the northeastern USA had emerged as the world's second major concentration of urbanized industry. The US population totalled 40 million, and the first transcontinental railroad had reached the west coast. In the next 30 years the population doubled and the trackage of railroad and volume of manufacturing output both quadrupled. As immigrants continued to pour into the eastern cities, and earlier settlers moved westward to open up the Great Plains for grain production, the industries of the Manufacturing Belt grew rapidly to keep pace with domestic demand for consumer goods and capital equipment (Meyer 1983). The 'gateway cities' of the Mid-West, such as Chicago and St Louis, became the centres through which produce from the expanding frontier was channelled to eastern and overseas markets (see p.45). On the basis of its coal deposits and early iron-ore workings, Pennsylvania developed as the raw-material hub of the industrial economy of steel and steam.

None of the other countries settled by Europeans in the late 19th and early 20th centuries had a population large enough to support domestic industry on the scale of the USA. In 1870, newly independent Canada sought to nurture manufacturing behind a protective tariff, but US industry had already benefitted from a similar policy for a century. Some of the first foreign branch plants of established US firms were prominent among the early Canadian producers. Canadian development continued to rely principally on the production of staple resources for export, with the US market dominant in forest products (timber, and later pulp and paper) and the British market in wheat.

The same dependence was evident even more starkly in the Southern Hemisphere, where settlement was so remote from the industrial economies on both sides of the North Atlantic (Denoon 1983). Large-scale immigration to Australia was stimulated by a gold rush in the 1850s, but sustained growth needed staples which could be marketed competitively in Europe. Wool, with a higher value-to-weight ratio even than wheat, and as easily transportable, was the 'crop' best suited to the country's environmental resource base and peripheral geographical location in the world economy. Argentina, where immigration helped to quadruple the population between 1870 and 1914, developed on a similar basis of wool, wheat, and also beef. With the exception of Canada, where US investment and sales of manufactures were significant, the overseas regions of large-scale 19th century European settlement were developed by European capital and equipped with European machinery. In both categories, despite the growing industrial capability of Germany and France, it was Britain which clearly dominated, as banker and 'workshop' of the world.

Trade and empire

The world-economy of Pax Britannica

By 1815 Great Britain had emerged unquestionably as a global hegemonic power. British industry was increasingly demonstrating its technical superiority over that of other nations, and the British armed forces had established their nation's military and political ascendancy over its principal European rival, France. Between 1815 and approximately 1870 the capitalist world-economy expanded in size and geographical reach under the relatively stable conditions guaranteed by British supremacy. Mercantilist policies were abandoned as counterproductive, for Britain's economy no longer needed protecting. Even British landowners survived the repeal of the Corn Laws, as growth in domestic food demand provided markets for their output in addition to imports. Logically, therefore, liberal (*laissez-faire*) capitalism became the foundation of British policy – or for most of it, for there was a significant exception. Trade rather than colonialism shaped British dealings with Latin America and Africa, but not with India. The deliberate destruction of the Indian textile industry in order to consolidate Lancashire's world leadership in cotton goods was a crucial anomaly in the otherwise 'home-grown' dynamism of Britain's Industrial Revolution.

India   The Treaty of Paris (1763) established British supremacy in the only two extensive areas of French colonization, North America and India. With the loss of the 13 North American colonies soon thereafter, Britain's attention turned increasingly to exploiting its Asian empire. Land taxation bore heavily on the native agrarian economy, being instituted in such a way that it destroyed traditional communal structures of land use and polarized rural society between rich, tax-collecting landlords and poor or landless peasants. Moreover, revenues were used to offset a 'public debt of India', notoriously inflated by expenditures doing little to benefit India but a lot to advance British strategic interests in Asia as a whole. The treatment of India's traditional manufacturing industry was devastating. Cottage textile production, similar to Britain's in the early 18th century, clothed a population of some 200 million and produced high-quality articles which were sold (by the British East India Company at a 100% profit) in Europe. But pressure from the mechanized Lancashire textile industry for protection from the competition of low-wage Indian craft producers persuaded the British government to enact tariff legislation which transformed the situation. The schedule of 1814 imposed a 3.5% duty on British cotton goods imported into India, but one of 70–80% on Indian cotton textiles entering Britain. With vastly increased scope for economies of scale, British producers were enabled to undercut Indian producers in their own market. Between 1814 and 1844 'the number of pieces of Indian cotton goods imported into Britain fell from 1.25 million to 63,000, while British cotton imports into India rose from less than 1 million yards to over 53 million yards' (Stavrianos 1981, p.247). The population of Dacca, 'the Manchester of India', shrank by 80%. The long-term repercussions became very evident during the remainder of the 19th century, for as the population of India rose, the number of people earning a living from manufacturing industry fell (Bagchi 1976).

Latin America and Africa These two continents were at markedly different stages of political development in the mid 19th century. By 1830, almost all of Latin America had achieved its political independence from the colonial rule of Spain or Portugal. Removal of the mercantilist constraints which had been imposed with diminishing effectiveness by the waning imperial powers left Latin American markets all the more open to British entrepreneurs. Unlike the USA, whose southern plantation owners' preference for free trade (which maximized the market for their crops) was overruled by Yankee industrialists bent on developing an indigenous tariff-protected manufacturing industry, the newly independent Latin American nations were generally governed by commercial interests (supporting free trade), who triumphed over the conservative landowners supporting domestically focused development (Stavrianos 1981). By 1914 Latin America accounted for 20% of the world total of long-term foreign investment, but this substantial dependence on imported capital for economic development introduced it at an early date to *neocolonialism*. Despite the achievement of political independence, control over, and the profits of, economic activity stayed predominantly in European hands.

Africa, in contrast, remained very much 'the dark continent' to Europeans. For a variety of environmental reasons (climate and disease, and the lack of large navigable rivers), European penetration outside of South Africa and Egypt was minimal, but contact with coastal peoples was extensive. The slave trade did nothing to promote the commercial development of West African resources, but as slavery was gradually renounced by Britain and then by other countries, 'legitimate trade' soon flourished in its wake. The Industrial Revolution created new demands for tropical products such as palm oil, used for soap and lubricants. Rather than being grown on European-owned plantations, many commercial crops were produced by native cultivators and their delivery to European traders was kept firmly in the hands of African middle men (Stavrianos 1981). This system of production was perfectly satisfactory to British free-traders: they obtained raw materials and marketed their manufactures within a stable set of commercial relations without the trouble and expense of colonial rule. Yet during the last quarter of the 19th century the situation changed radically, as the European hold on the global economy was tightened in a flurry of imperial expansion, not least the 'scramble for Africa'.

The global economy completed

Between approximately 1870 and 1914 the Europe-centred capitalist world-economy penetrated almost the entire globe. Britain's hegemony remained intact, but there were growing signs that its economy was no longer the pace-setter. The USA overtook Britain in the volume of its industrial output, and also in the sophistication of its mechanical engineering and the organization of mass production. In Europe, a united Germany increasingly exploited its central location in the dense continental railway network to capture new markets, and challenged British manufacturers worldwide with innovative products, derived from a strong emphasis on applied scientific research. The global core of urbanized industrial societies was thus becoming larger and more differentiated internally. At the other end of the spectrum of economic development, all those regions of the world (with two notable exceptions) which had so far been neglected by Europeans were finally brought within their political and economic orbit.

Imperialism reaches its zenith Having lost its first empire to Britain, France pieced together a second one after 1830, this time in north-west Africa and South-East Asia. Britain, too, continued to acquire new colonies, many of which were strategic points on the major oceanic trade routes, especially those serving India. The pace of imperial expansion accelerated markedly after 1870, however, primarily reflecting the emergence of newly unified nations such as Germany and Italy (and even the USA), intent on securing for themselves control over territories not yet held by existing imperial powers. With the exception of Holland, which remained satisfied with its long-held and profitable colonies in the East Indies (Indonesia), the other imperial European powers joined in the rapid and final carving-up of the world's landmass. Africa in particular was fragmented into territorial units which totally ignored the boundaries of native political systems (Meinig 1969). Only in China, where a weak central government was still strong enough (and sufficiently remote from Europe) to limit imperial penetration to trading enclaves, and in Japan, where a strong nationalistic government and culture kept European influence at bay, was the wave of colonialism successfully deflected.

What prompted this dramatic finale to the expansion of the European world-economy? It is too often explained in simplistic abstractions, such as 'the imperative of capital accumulation' or the 'irresistible march

of progress'. Imperialism embodied a whole range of motivations by
the peoples of the dominant powers – economic advantage, strategic
security, cultural supremacy, and religious conversion – and they became
so intermixed that ideological rationalizations could be produced for
any and every action. Today, we are all too conscious of the negative
impacts of imperialism, which makes it difficult to evaluate the positive
achievements of the colonial powers in perspective. Within the narrower
field of economic explanations, Wrigley (1978) has attempted to account
for the renewed acceptability of colonialism in late-19th century Britain,
when in the middle of the century the free-traders had been glad to
demonstrate that it was an unnecessary burden. Pressure to return to
a policy of (neo)mercantilism came (despite Marxian claims to the
contrary) not from industrialists desperate for captive markets, but
from trading groups wanting to preserve local monopolies overseas,
and from statesmen who, still coming largely from the landowning
élite of a small island, were convinced of the long-term importance of
controlling natural resources. And in an age when overland accessibility
depended on railways, control required formal political annexation to
ensure peace and orderly government in the territories through which
the trains ran.

The multilateral trading system   By the early 20th century, commercial
linkages pioneered by Europeans had been expanded and consolidated
into an integrated capitalist world-economy. Britain retained its key rôle
in financing and transporting the world's commodity trade, but the
nature of the commodities which were shipped, and the geographical
pattern of their movements, reflected the enlargement of the core
of industrialized countries and the increasing economic specialization
of many parts of the less developed periphery. International trade
could not have maintained its steady growth without continuous
improvements in transportation technology, the development of an
international telegraph network, and the radical changes in international
accessibility brought about by the construction of the Suez and Panama
canals (Latham 1978; Fig. 4.9).

   The pattern of international merchandise trade balances (Fig. 4.10)
indicates how interlocked the global trading system was. Of the five
regions identified, two had large deficits in their *visible* (merchandise)
trade. Britain had been a net importer of goods since 1816, and as
the Industrial Revolution changed the face of continental Europe,
that region also became a net importer. Despite the success of British
and continental manufacturers in selling their products throughout the
world, the rising standard of living of Europeans and the resource needs
of their industries kept imports running ahead of exports. The USA, with
a much more favourable population/ resource ratio, was able to support

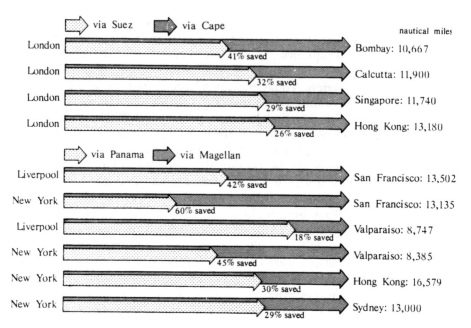

Figure 4.9 Impact of the opening of the Suez and Panama Canals on international shipping routes.
Source: based with permission on material published in 1982. *The Times concise atlas of world history*. London: Times Books, p.109.

its economic growth with a net deficit only with the underdeveloped countries whose tropical resources it lacked. Exports of machinery were prominent in its visible trade surpluses with Britain, Europe, and the lands of European settlement (notably Canada). The less developed countries, as a group, amassed a visible trade surplus almost equal to that of the USA, exporting raw materials on a vastly greater scale than they imported manufactures. Only Britain maintained an export surplus in its trade with them.

This pattern of visible trade balances immediately prompts some questions. How did Britain and Europe cover their deficits? Why did the surplus of the USA promote its development when that of the less developed countries appeared not to promote theirs? Part of the answer lies in the contrasted pattern of *invisible* earnings from services such as shipping and banking. In 1928, the visible trade surplus of the underdeveloped countries ($1130 million) was more than eaten up by their payments for shipping alone ($1143 million), which almost entirely benefited British and other European carriers (Frank 1979). Another part of the answer lies in the rôle which Britain made India play in the system of international financial settlements by forcing it to remain a captive market for British textiles. This merchandise trade left the Asian region with a substantial deficit, which it more than covered, however, by the surplus earned on merchandise trade with Europe, the USA, and other parts of Asia. Saul (1960) estimates that Indian trade alone financed about 40% of Britain's total deficits. Through its control of the Indian economy, Britain was able to survive the growing protectionism by which, at the close of the 19th century, newly industrializing countries attempted to stimulate domestic manufacturing at the expense of British exports. Without the substantial visible surpluses earned by its tropical colonies in India, Malaya, and West Africa, Britain would not have been able to settle its trade deficits with the developing (and tariff-protected) industrial economies of Europe and North America. In turn, its demand for goods from these regions played an important part in promoting their economic growth (Latham 1978).

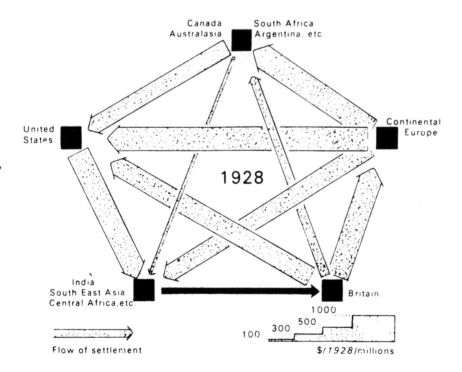

Figure 4.10 Multilateral balances of merchandise trade in the early 20th century. Settlements flow from regions with a trade deficit.
Source: Latham, A.J.H. 1981. *The depression and the developing world, 1914–1939*. London: Croom Helm; Diagram 1.

Conclusion

The years around 1920 marked the end of an era in many significant ways, even if they did not all become apparent immediately. The USSR severed most of its links with the capitalist world and embarked on the first attempt ever to organize a national economy following Marxist principles (see Ch. 9). The leading European nations were materially and psychologically exhausted by the First World War. Britain's hegemony had been decisively undermined, but strong isolationist sentiments in the USA made that nation disinclined to exert the global influence of which it was increasingly capable. The boom of the third Kondratieff wave (Fig. 1.1) soon gave way to the most traumatic slump in the history of industrial capitalism (see p.102). In many respects, the evolution of the world-economy marked time until growth resumed after the Second World War.

Understanding the past is crucial to interpreting the present. Despite dramatic changes in the volume and structure of world trade in recent decades, there are substantial continuities between the developments recorded in this chapter and the geography and dynamics of the contemporary world-economy. These can be demonstrated in three different spheres. First, we have identified the origins of the division of the modern world into an economically developed core and a less developed periphery. Secondly, we have noted the practical implications of environmental and spatial attributes for the course of social and economic development. Thirdly, we have noted the influence of societies' philosophical and ideological perspectives on the character of development in different parts of the world.

5 States and the global economic system

Of the two principal groups of actors in the contemporary global economy, states and transnational corporations, states are significantly more diversified and their behaviour more complex. It is therefore necessary to analyse their characteristics in some detail to understand the varied ways in which they interact with the world-economy and shape the economic geography of their own territories. One identifiable subgroup, the industrialized state-socialist nations, the integration of which into the capitalist world-economy is only partial, is analysed separately in Chapter 9. In this chapter, the remainder are grouped into two sets: the Western liberal democracies (the industrialized capitalist nations), and less developed countries, the independent statehood of which is in most cases of recent origin. This division is obviously oversimplified, for states differ across a range of characteristics rather than fall unambiguously into discrete classes. But there are enough commonalities within each set to justify the classification.

The Western, industrialized 'capitalist state'

The capitalist state

Capitalism emerged as a mode of production in a specifically European cultural context. The gradual breakdown of the feudal system involved the dissolution of a redistributive, hierarchical society. It was replaced by a society of pervasive market-exchange relationships between relatively autonomous individuals. Governments representing the economically dominant social groups (mercantile entrepreneurs gradually replacing landowners) presented themselves as the legitimate foci for popular allegiance and began to act with increasing directness, both internationally (mercantilism) and domestically, to foster national economic development. The Industrial Revolution in Britain was primarily the outcome of individual initiative within a supportive cultural and political environment, but in nations which followed Britain's path, such as the USA, Canada, France, and Germany, state authorities took a more active rôle in establishing an industrial economy, through tariff policy, infrastructure investment, etc. (see Ch. 4).

The political economy of early 19th-century industrial capitalism was based, as Marx perceived, on a class structure generating conflict between a minority of capitalists and the majority of wage-dependent labourers. It also rested on a fundamental alignment of interests between the state authorities and the capitalist class. Marx saw the resolution of this conflict as lying in a socialist revolution which would abolish the class structure and the antagonism between those in power and the mass of the people. However, the socialist states which have subsequently come into being embody their own distinctive forms of social tension and alienation (see Ch. 9). Moreover, alternative forms of social change within capitalism proved possible.

Giddens (1981) distinguishes between the general question of the rôle and functions of 'the state in capitalism' and the narrower issue of the nature of 'the capitalist state'. The latter term denotes the specifically European cultural and political characteristics of the state as it has evolved in the core nations of the West. He summarizes the basic tension facing 'the state in capitalism' in the following terms. In contrast to previous political systems,

> The accumulation process in capitalist societies . . . rests upon the mobilization of privately owned capital, and . . . is not under the direct control of the state. At the same time, the state assumes responsibility for the provision of a range of community services derived from state revenues which depend upon the 'economic success' of the economic activities of employers and workers (Giddens 1981, p.214).

The distinctive character of 'the capitalist state' arises out of the way in which that tension has been not resolved, but at least contained. For, to a degree that Marx did not anticipate, the working class of early industrial capitalism secured the legal right to organize to defend its members' interests, first in the workplace (by the formation of trades unions), and subsequently in national legislatures (when the extension of the franchise opened the way for the formation of political parties representing the labour movement). These rights, and widespread acknowledgement of their legitimacy, were not won without a struggle, but the fact that they *were* won in the liberal democracies of the West was decisive in steering the political economy of these nations away from unfettered capitalism. A form of 'mixed' capitalism emerged, in which the State (the capital 'S' signifying the state as an increasingly powerful and pervasive social

institution) began to play a notably expanded rôle in regulating social
and economic life.

As a result of this evolution, the 'capitalist state' possesses a 'relative
autonomy' in relation to the principal economic institutions of capitalist
society. Rather than acting simply to further the economic interests of
the capitalist ruling class, as Marx and Engels claimed in *The communist
manifesto* (1848), 'the state is best seen as a set of collectivities concerned
with the institutionalised organisation of political power' (Giddens 1981,
p.220). The policies of the State reflect the outcome of the dynamic
conflicts of interest in contemporary society. These involve not only
capitalist owners of the means of production (encountered more often
as a corporate managerial bureaucracy or the managers of pension funds)
and their wage-dependent employees, but other politically powerful
social groups, not least the numerous employees of the modern State
itself: for the 20th century has been marked by a dramatic growth in the
functions of the State, operating at various geographical scales, in ways
which are examined more closely below.

The evolution of the capitalist state
Adam Smith had restricted the legitimate activities of the State to
national defence, the administration of justice, and the provision of
public infrastructure and institutions (O'Brien 1975). However, the
development of an urbanized industrial economy soon made it clear
that the State would have to assume a much wider set of responsibilities
than it had in the past. The expanded rôle of government in a capitalist
society involved four distinct but overlapping functions:

(i)   'supplier' of public or social goods and services;
(ii) 'regulator and facilitator' of the operation of the market place;
(iii) 'social engineer', in the sense of intervening in the economy to
      achieve its own policy objectives; and
(iv) 'arbitrator' between competing social groups or classes (Clark &
      Dear 1981, p.49).

For example, the densely settled urban environment required physical
infrastructures to provide essential services for which individual capitalists accepted no responsibility. These called for public institutions
responsive to the civic priorities of industrial and commercial interests.
Public health was an early focus of intervention: municipal authorities
were frequently in the position of *providing* the infrastructure for
water supply and sewage disposal, *regulating* the residential construction
industry to curb the speculative building of insanitary dwellings, *engineering* desired standards of urban amenity through planning legislation,
and continually *arbitrating* disputes between social groups which were
differentially affected by these actions (Briggs 1968). Harvey (1985a)

argues that the more the politics of large cities became coloured by working-class priorities in the late 19th century, the more bourgeois (capitalist) objectives were achieved through the policies of higher levels of government.

The growing rôle of the 'capitalist state' can be interpreted in a wider perspective as a necessary social and economic response to the conditions of early capitalist industrialism. Legislation to protect and promote the welfare of every citizen gradually accumulated in an environment shaped by the increasing strength of the labour movement (more evident in Britain than in North America); the better articulated ethical critique of humanitarians and the churches; and the ruling classes' own recognition that to allow the poverty and insecurity of working-class life to persist invited revolutionary action. By the time of the First World War, public provision of at least elementary education, limited involvement in health-care delivery, and, in a few countries, the beginnings of a social security system had come into being. The war itself was significant for vastly expanding the State's experience of mobilizing the whole range of national economic resources, and for justifying a substantial increase in the level and scope of its powers of taxation.

Depression, Keynes, and the fruits of prosperity

The Great Depression of the 1930s was not the first Kondratieff downturn in the capitalist world-economy (see Fig. 1.1), but it was certainly the most severe in the scale of its human impact. 'In many industrial countries, over a quarter of the labour force was thrown out of work. Industrial production fell to 53 percent of its 1929 level in Germany and the United States' (Barraclough 1978, p.266). The extent of the decline in production, employment, and the value of fixed assets challenged belief in the effectiveness and political legitimacy of the capitalist system, but also, more immediately, in the adequacy of conventional neoclassical theory to serve as a basis for public policy:

> Government reactions to the Depression were unenlightened . . .
> Soon . . . the dire consequences led one government after another
> to attempt to stimulate the domestic economy . . . Even the most
> economically conservative regimes were forced to abandon *laissez
> faire* in favour of some degree of state intervention (Barraclough
> 1978, p.266).

The theoretical demonstration that such unorthodox government behaviour was precisely the correct response to the Depression came with the publication of Keynes' (1936) foundational statement of the macroeconomic dynamics of the capitalist economic system. By showing that there was no guarantee that a society's savings would be allocated to productive (and job-creating) investment, and that in a slump savings

are, in fact, whittled away to maintain necessary consumption, he demonstrated the rationale for government investment to restore consumer
purchasing power, business confidence, and thus the expansionary cycle
of economic growth.

The pursuit of Keynsian policies of economic management in the
Western core states for three decades following the Second World
War was a major contributor to the rapidly rising prosperity which
these nations enjoyed. By acting to counter the short-term business
cycles of the capitalist economy, governments were able to ensure full
employment and the continuous utilization of industrial capacity. But
other favourable trends, which Denison (1967) classified as increases
in the absolute level of available factors of production, and increases
in the efficiency with which these factors were combined and utilized,
contributed significantly to this economic growth. Rates of population
growth increased sharply after the years of depression and war, with
new family formation and the 'baby boom' stimulating residential
construction, production of consumer goods, and the expansion of
health and education services. Substantial defence spending, new capital
investment and, in countries such as Canada and Australia, the exploitation of newly developed natural resources, added to the *quantitative*
dimension of growth. However, *qualitative* changes, such as increases in
factor productivity stemming from technological advance, improved
managerial capabilities, economies of scale, and the redeployment of
labour from agriculture into other sectors of the economy were even
more important, except in the USA, which had already experienced
many of their benefits (Berry, Conklin & Ray 1976).

In a climate of steady economic growth, governments dramatically
increased the proportion of gross national product (GNP) represented by
their own spending; on the social services of the welfare state, defence,
and their participation, directly or indirectly, in goods production.
Differences between countries in the scope of State activity reflected
particular histories of the success of labour-based democratic socialist
parties in enlarging the *social wage* (guaranteed entitlements to social
services), and in incorporating specific industries within the public
sector (see Fig. 5.1). Dicken & Lloyd (1981) note that, by the mid-1970s,
government expenditure accounted for 32% of US GNP, and 56%
(including expenditures by public corporations) of British GNP. State
spending of this magnitude has played a significant rôle in shaping the
contemporary economic geography of Western nations (see Ch. 8).

The capitalist state in retreat?
The Kondratieff downswing which intensified in the early 1980s brought
a renewed awareness of the limits and ambiguities of extensive State
penetration into an economy still fundamentally driven by the pursuit

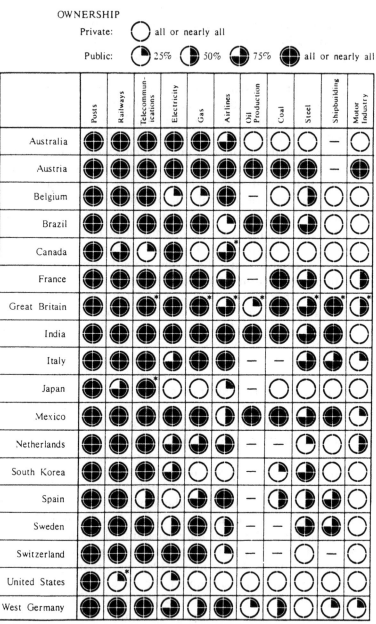

Figure 5.1 State ownership of selected industrial sectors, 1978–88.
Source: adapted from *The Economist* 30 December 1978, p. 39; updated from *Financial Times* (10 December 1986) and press reports.

of private profit. The 'capitalist state' remains ultimately dependent upon the prosperity of the national economy, which increasingly means upon its international competitiveness. After 1945, Western governments felt politically compelled to ensure rising real wages and the maintenance of an extensive welfare state. In a time of rapid economic growth, these obligations could be fulfilled without imposing excessive burdens on the process of private capital accumulation. But when, for a variety of reasons, growth became more elusive, yet public expectations of the State remained undiminished (and even increased because of rising unemployment, etc.), governments at all levels began to face a fiscal crisis (O'Connor 1973). Obligations previously accepted as politically legitimate have tended to be disavowed as the ability to finance them is weakened (see Ch. 12).

Many Western governments reversed their policy direction in the 1980s, adopting less interventionist ('monetarist') financial policies and a variety of measures with a pro-business thrust. Curtailment of welfare-state expenditures was explained as a 'necessary' adjustment to the difficulties facing their nations' corporate sector (in which average profitability had been steadily declining for two decades) and to their own unbalanced budgets. If the erosion of the welfare state is 'necessary', however, does it follow that so too is a more polarized society, similar to that which characterized industrial capitalism before the interests of wage earners found expression in the policies of the 'capitalist state'? By the late 1980s, considerable evidence had accumulated to suggest that social and economic disparities were widening: witness the growing North-South divide in the UK and the stagnant or declining living standards of many American workers.

Capitalism, socialism, and nationalisms

Introduction

The evolution of the capitalist world-economy was shaped by Europeans, whose influence derived from the conjunction of economic and political interests within a few core nation-states. That core expanded in the late 18th and 19th centuries as other politically unified nation-states, notably Germany and Italy, emerged, and as societies of European origin, including the USA and Canada, became similar independent nations. In every case, the achievement of 'national ambitions' (the articulated interests of an influential *subset* of the population) involved policies to strengthen the domestic economy in the face of competition from products of the more advanced British economy. Whether in the form of protective tariffs, or direct financial support for industry and related infrastructure (such as railways), national governments intervened in 'the development process'. Russia and Japan embarked upon comparable, but distinctive, programmes of 'nation building'.

In contrast, most of the Third World was, by the close of the 19th century, made up of politically dependent colonies, governed by expatriate civil servants for the benefit of the European imperial powers. As we noted specifically in the case of India (see Ch. 4), without the autonomy and institutions of sovereignty to protect their domestic economy, Third World societies witnessed what Frank (1966) has termed 'the development of underdevelopment'. Yet, significantly, this phrase was coined in the context of Latin American experience. For despite the achievement of political independence by 1830, in these societies the interests of the locally dominant élites coincided with policies designed to maintain external control of the national economy by institutions based in the global core.

Two points arise from this brief review. One is that the powers and institutions of statehood are critical parameters in the process of development, which involves more than simply 'economic development' (see Ch. 10). Indeed, the political and economic dimensions of development, insofar as they can be meaningfully distinguished, are structurally interrelated. The dynamic tension which characterizes relations between sovereign states and supranational economic institutions or TNCs in the contemporary capitalist world-economy points to the importance of nationalism as a factor influencing State action in countries whose independence is relatively recent and often fragile.

The second point, however, is that to regard 'nations' or 'nationalisms' as homogeneous categories is seriously misleading. To possess a seat in the United Nations General Assembly may satisfy one definition of what constitutes a sovereign state, but there are major *qualitative* differences between, say, the Belgian State (to choose a second-rank member of the European core) and the Nigerian State. Similarly, the cultural character and political constituency of 'nationalist' policies varies considerably between states, so that the potential for mobilizing popular support for the social and economic changes which development entails is far from uniform. The European-type 'capitalist state' must be recognized as a special case rather than the norm: it cannot be assumed to represent a prototype of the State in Third World nations (Giddens 1981, Sandbrook 1986). Explicit acknowledgement of the rôle of sovereign states in the development process must be accompanied, therefore, by interpretation of the actions of specific states in their own context.

Different states: different routes to development
Seers (1979b) provides a simple 'ideological map' for classifying the character of national regimes and the strategies of development which

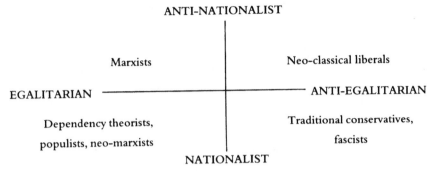

ANTI-NATIONALIST

Marxists   Neo-classical liberals

EGALITARIAN ———————————————————— ANTI-EGALITARIAN

Dependency theorists,   Traditional conservatives,

populists, neo-marxists   fascists

NATIONALIST

Figure 5.2   A contemporary ideological map.
Source: Seers, D. 1979b. *The congruence of Marxism and other neo-classical doctrines,* Discussion Paper DP 136, Institute of Development Studies, University of Sussex; p.16.

they implement (Fig. 5.2). He combines a capitalist–socialist axis, defining the basic structure of a state's political economy, with a nationalist–anti-nationalist axis, defining the orientation of a society to the world at large. The top right quadrant represents the philosophy of hegemonic liberal capitalism, such as was embodied in the policies of Britain in the mid-19th century and of the USA in the mid-20th. Elaborated on the basis of neoclassical economic theory, this outlook is distinctively associated with a belief in the efficiency and effectiveness of market mechanisms. It therefore argues strongly for restricting State 'interference' in the economy to an absolute minimum: it encourages free trade and the international movement of factors of production in externally oriented policies, and discourages substantial interpersonal or interregional redistribution of income domestically.

The top left quadrant is best represented in the writings of Marx, for his analysis of the evolution of the global economy led him to expound a philosophy which was explicitly socialist and universalist. The call of *The communist manifesto,* 'Workers of the world, unite!', was based on the firm belief that individuals' class interests would far outweigh their nationalist sentiments. This expectation has been disappointed, not only in the capitalist core, where states have acted to promote the welfare of workers (not least by protecting their jobs against the competition of low-wage producers in the Third World) but also in the socialist core, where Soviet policies towards the employment of Vietnamese 'guest workers' or trade with other socialist states reflect national interests more strongly than concern for class solidarity.

The bottom left quadrant represents the outlook of states which interpret their experience of incorporation into the world-economy in terms of *dependency theory.* More accurately, this is a school of theories, among which there are important distinctions (see. Ch. 9). This quadrant is the polar opposite of the neoclassical liberalism espoused by hegemonic powers: the rôle of the State lies in eliminating, as far as possible, the penetration of external capitalist institutions and in becoming the primary agent of indigenous economic development. Note that in being 'nationalist', this position is, strictly speaking, *neo*-Marxist. Its interpretation of capitalist development's rôle in bringing about the necessary conditions for socialism is essentially negative, whereas Marx's optimistic assessment of capitalism's effects is shared today by a minority of 'radical' commentators on the contemporary Third World (Warren 1979).

The bottom right quadrant represents the philosophy of states which amply substantiate pessimistic views of capitalist development. Characterized by repressive and authoritarian governments (typically a military dictatorship), LDCs in this category protect their low-productivity economies from foreign competition. At the same time, however, they preserve the monopoly profits and/or rents of the few resource-based TNCs on which their export earnings depend, and the power and wealth of the small domestic élite of large landowners and commercial interests (the 'banana republic' pattern).

Obviously, not only do individual states differ in their location on this ideological map, but the position of each one changes over time, sometimes suddenly. (The same is true of individual theorists, so the map is intended as a guide to *interpreting* the stance of states and scholars, not a means of 'labelling' them.)

Third World states in 'world time'
Material aspects   We noted in Chapter 4 that the Industrial Revolution in Britain was a fundamentally *evolutionary* process, and one which pre-dated the transformation of manufacturing into a science-based activity. It was possible for individuals with a rudimentary level of literacy and numeracy (and often without it) to advance the efficiency of industrial production by simple improvements of familiar technologies. 'Learning by doing' was a feasible means of entry into most branches of employment. However, the newly independent LDCs of the 1950s and 1960s found themselves in a world in which the basic processes of transforming natural resources into useful end-products were embedded in technologies incorporating areas of knowledge far beyond the immediate reach of their populations. Not only were the *preconditions* of sustained economic development which existed in Britain (rising agricultural productivity, accumulated capital stock, educational and cultural attributes) lacking in these nations, but the transformation of their economies involved bridging a discontinuity between 'traditional' and 'modern' methods and outlooks enormously greater than that experienced in 19th-century Europe. Moreover, science-based technology is embodied in institutions of the global

core, such as TNCs, which regulate its diffusion in ways that are compatible with the maintenance of their global leadership. Any Third World country which aspires to 'catch up' to the industrialized core nations in terms of the efficiency and diversity of its economic activity faces severe handicaps: the problems of starting its development process 'late' in 'world time' (Giddens 1981) are ones which a contemporary LDC cannot escape. Even the potential advantage of a latecomer, being able to draw upon the accumulated scientific and technical knowledge of the developed nations, has its dangers (Myrdal 1957). This knowledge needs to be critically evaluated and adapted to the specific conditions and goals of the developing society.

Cultural and political aspects Latecomers to the development process also encounter political problems of an entirely different character from those faced by the core nations in the early 19th century. Hirsch (1976a) points out that the political economy of Western, market-based, liberal democracies, such as the UK and the USA, embodies implicit promises which became realized in society only over a considerable period of time and in a specific sequence. The *bourgeois revolutions*, in which the commercially minded middle classes took over political power from an aristocratic élite, succeeded in gathering wide support because their ideologies suggested that they would bring about civil liberties, political freedom, and economic opportunity for every citizen. In practice, civil liberties became guaranteed (though not for blacks in the USA) long before the political rights of freedom of association (especially for labour unions) and of 'one man, one vote' were finally conceded (and women had to wait even longer). Meanwhile, throughout the 19th and early 20th centuries, economic development took place in Western Europe and North America in societies which did little to provide equality of economic opportunity, or guarantee freedom from abject poverty. Only when the universal franchise enabled working-class populations to pursue their economic interests in the political arena was progress made toward establishing the more egalitarian economic conditions associated with the welfare state. Not until the core nations had achieved a high level of economic development (as measured by GNP per capita) did the interpersonal redistribution of income which this requires become a politically irresistible expectation of the State.

For good or ill, the Third World is the heir of the world's dominant, Western culture (Ellul 1978). The concept of individual freedom, which has been tortuously and imperfectly woven into the fabric of the liberal democracies (and so far subordinated to authoritarian central institutions in existing socialist societies), also nourished the aspirations for freedom from colonial rule and for the political independence of 'sovereign peoples' which, after 1945, brought the contemporary Third World into existence. The problems which this raises for a nation's pursuit of economic development and the consolidation of State institutions are threefold.

First, although the impact of moral suasion should not be underestimated in accounting for the rapid retreat of colonialism after 1945, political independence rarely came without conflict, whether between the colonized and the colonizers or between rival claimants to domestic political legitimacy. The material destruction and economic dislocation which this entailed and its frequent legacy of communal distrust has handicapped progress towards national integration and economic development.

Secondly, LDC governments have begun their efforts to raise the standard of living of their populations in an era of 'world time' when the expectation that the State will act to secure the welfare of all citizens, especially the poor, has much more political salience than it did in core nations at a comparable stage of development. The scope for interpersonal income redistribution on a scale that would tangibly improve the conditions of economically disadvantaged groups is much less in countries that are already characterized by meagre average incomes, however unequally distributed, than in rich core states. The challenge of achieving economic growth while limiting the political disaffection arising from unfulfilled expectations makes the management of Third World economies very difficult. The degree to which growth and equity are compatible development goals is discussed more fully in Chapter 11.

Thirdly, the efficient functioning of economic institutions in many LDCs is hampered by the fact that Western culture is alien to large sections of the population and its values have been imposed on society 'from above' rather than fought for 'from below' as they were in Europe. In rural India, for instance, village leaders who uphold traditional values which give little place to a person's independence of thought and action (especially a woman's) often frustrate development strategies which are designed to appeal to individual or collective initiatives.

Society and the State

Unlike the core nation–states of Europe, in which people of a common culture achieved political control of a unified territory early in modern history, and unlike the USA where diverse streams of European settlers in a common territory subscribed to the unifying ideology of 'the melting pot', many Third World states, especially the 'young' ones of Africa, are still struggling to define a national identity which will transcend their deep internal cultural divisions. The colonial boundaries drawn on maps in London, Paris, or Berlin, which almost totally ignored the human and physical geography of the territories they parcelled out, are the immediate source of the cultural heterogeneity of so many LDCs. Note, however, that very few of the tribal or linguistic groups of indigenous Third World peoples possessed the territorial basis for a viable 'European-type' nation–state. Rivalries and antagonisms between tribes or other social groups which, during the colonial era, were checked in some places and exacerbated in others by the imperial power, have in many countries been resurgent since Independence. Creation of an effective, professional, national administration and the implementation of national economic policies (which are bound to affect some people's interests adversely) is very difficult so long as society is fragmented by inter-group mistrust, violence, and threats of secession (Nafzinger 1983).

This source of social division intersects, with complex results, with that based more directly on socio-economic stratification. The two often come together in what Myrdal has termed 'the soft state':

> The concept of *the soft state* identifies: a general lack of social discipline in underdeveloped countries, signified by deficiencies in their legislation and, in particular, in law observance and enforcement, lack of obedience to rules and directives handed down to public officials on various levels, often collusion of these officials with powerful persons or groups of persons whose conduct they should regulate, and, at bottom, a general inclination of people in all strata to resist public controls and their implementation (Myrdal 1970, p.229).

Corruption, which promotes arbitrariness, obstruction, and delay in the machinery of government, is also part of this phenomenon (Sandbrook 1986). As Myrdal emphasizes, to identify these problems is not to engage in moralistic analysis, as if core nations were free of pork-barrelling or bribery, but to face the fact that development is much more difficult to achieve in societies in which a lack of social discipline is pervasive.

Weinstein (1976) demonstrates that the failure of LDCs to maximize the potential national benefits of foreign direct investment often results as much from the capriciousness and ineffectiveness of State administration as from deliberately obstructive behaviour on the part of TNCs, although he does conclude (1976, p.403) that the latter 'help to keep the soft states soft'. Where corruption is allied to despotic rule, the extent of 'anti-development' is even more pronounced. For instance, an internal World Bank report on Haiti claimed that in 1975 more than half of the country's public revenue was 'credited directly to 300 or so special accounts [in the National Bank] for a multitude of unspecified purposes' over which the Finance Ministry had no control (*Guardian*

Weekly 5 June 1977). The fall of the Marcos regime in the Philippines in 1986 similarly revealed misappropriation on a grand scale.

The quality of public administration is so significant in the Third World precisely because of the very wide range of responsibilities which the State has assumed in economic affairs. The degree to which successful rural development, as indicated by improved agricultural productivity and social welfare, has been achieved in contemporary LDCs appears to be strongly associated with local institutions which are *accountable*; and to perform effectively, these require a commitment to the same principles at higher levels of administration (Uphoff & Esman 1978).

Conclusion

Governments around the world, including those of state-socialist nations, differ significantly in their political base, their popular legitimacy, their administrative competence and their policies towards engagement in the capitalist world-economy. The common characteristics of 'capitalist states' derive from inherited cultural values which have ensured that the State is more or less responsive to the interests of the majority of citizens and not only to those with economic power. Nevertheless, the State's dependence on the success of capitalist enterprise for its revenues and its ultimate legitimacy leads to continuous tensions between competing social interests. These can be accommodated more readily during periods of economic growth than of stagnation or decline.

Third World states are much more diverse than those in the global core, but they are uniformly poor in comparison (with a few oil-exporting exceptions) and their governments are weaker as economic agents. Much of this weakness stems from a history of subordination to core interests, who generally imposed a territorial definition of the state, shaped its economy in their own interests, stifled indigenous political and economic initiatives, and who continue to exercise economic dominance even in the post-colonial era. Contemporary problems of development are magnified by the passage of 'world time', which presents the State with challenges not faced by governments at the start of the industrial era. Finally, the cultural matrix within which the State is defined and in which it must operate is in many cases an obstacle to effective action.

Yet, wherever it lies on Seers' 'ideological map' (Fig. 5.2) and whatever its strength or weakness as an institution, the modern State is the formal arbiter of all activity that takes place within its territory. It influences, directly and indirectly, the geography of its domestic economy; and it mediates, to an extent that its effective authority and political inclination define, the interaction between that economy and the wider world. Economic actors in the capitalist world-economy have greater freedom to manoeuvre than a territorially bounded state (see Ch. 1), but their operations cannot be stateless. Strategies of engagement between states and the dominant business corporations of today are reviewed in Chapter 7.

6 The corporation and the global production system

The structure of production

The production of marketed goods and services takes place in millions of workplaces around the world. Whether they are farms or offices, mines or factories, these 'establishments' are integrated into production systems which satisfy the great variety of human needs. To understand the characteristics of each one (its size, location, output, employment, the technology of production, etc.), we need some basis for tracing its relationships to the wider, and ultimately the global, economic system. This chapter identifies the 'corporation', the institution which exercises ownership and control of production establishments, as the key intermediary between the workplace and the world-economy. There are, of course, myriads of small businesses in which the 'corporation' (or 'firm') is no bigger than the 'establishment': most commercial farms fall into this category. But large firms, which are *multilocational* (they operate many different establishments), and increasingly *transnational* (they do business in many different countries), dominate the structure of the contemporary economy.

Even the largest TNCs, however, operate within a dynamic system that they do not control. They enjoy a geographical flexibility in responding to changing economic conditions which gives them leverage in dealing with national governments, but they are bound by the laws and regulations (to the extent that they are enforced) of the countries in which they seek to do business. Moreover, despite their power and resources, they are not immune from competition, whether from similar firms or from smaller ones specializing in particular product lines. Indeed, the increased turbulence of the capitalist world-economy since the early 1970s has profoundly altered the ways in which firms of all sizes organize their business and respond to market signals. Therefore, although the focus of this chapter is the corporation, we also need to analyze the overall production system which is its environment.

Territorial concentration
The global output of goods and services is concentrated within specific geographical regions. On the basis of gross domestic product (GDP) measures, approximately two-thirds of total world production in the mid-1980s took place in the industrialized market economies of the West, with the balance split almost equally between the state-socialist economies (including China) and the less developed countries (which are nearly all integrated into the capitalist world-economy). Within each of these three broad groups of nations, production was further concentrated in the leading economies such as, respectively, the USA and Japan, the USSR and China, and Brazil, India, and Mexico (World Bank 1986). When per capita indices are used as measures of comparative productive capability, the globally dominant position of the industrialized nations of the capitalist core emerges even more starkly. World Bank (1988, table 1) data reveal that in 1986 the weighted average gross national product (GNP) per head of 39 'low-income countries' was $270, whereas that of 58 'middle-income countries' was $1270, while that of the 19 'industrial market economies' was $12 960.

Sectoral concentration
The process of economic development involves a society which is initially almost entirely dependent on production within the agricultural sector diversifying its allocation of human and capital resources into other forms of activity. The contrasts in the sectoral composition of some contemporary national economies are illustrated in Figure 6.1. The historical experience of the capitalist core states has been for manufacturing's share of output to grow as that of agriculture declines; then in the post-1945 period the rapid expansion of the service sector has outpaced growth in manufacturing, causing that sector's share to decline. For reasons which are examined in Chapter 11, the growth of most less-developed nations has been marked by the weak emergence of manufacturing.

Within the non-agricultural sectors especially, the structure of output has changed significantly over time, reflecting broad changes in technology and social organization. The 'coal, steam, and iron' technology of the 19th century gave way to 'oil, auto, and steel' technology in the first half of the 20th century (see Fig. 4.4); and today we can point in addition to the significance of petrochemicals and microelectronics in the composition of industrial output. The service

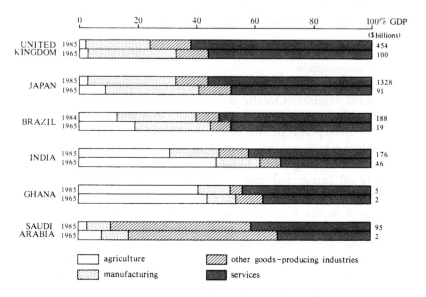

Figure 6.1 Sectoral composition of gross domestic product in selected countries, 1965 and 1985.
Source: compiled from World Bank 1987. *World development report 1987*; Annex Table 3.

sector has also been transformed since the middle of the 19th century, when numerically it was dominated by domestic personal servants, whereas today it is characterized by professionals in health, education, and business, and the staff of government agencies and commercial services (Gershuny 1978).

Corporate concentration
The ownership and control of production capacity in the non-socialist world also exhibits strong concentration. Even in agriculture, the sector most characterized by large numbers of independently owned establishments, there is a marked tendency in the industrialized nations for marketed output to originate from a small proportion of them. In the non-agricultural sectors, the extent to which global output is controlled by a few corporations who own a minute fraction of the world's production establishments is very dramatic. General Motors and Ford together produced over 37% of the non-communist world's motor vehicles in 1979, and together with only six other firms they were responsible for three-quarters of total output (Maxcy 1981). Despite the growing tendency for oil-exporting nations to nationalize their petroleum industry, the seven major oil TNCs were still responsible for handling 43% of all crude oil produced in the West outside North

America in 1980 (UN 1983b). This pattern is repeated in varying degrees in other sectors, where one or two firms provide a substantial proportion of total output: one thinks of IBM in mainframe computers and Boeing in commercial airliners.

This corporate concentration of production reflects a number of underlying pressures within a capitalist economy. One is that although the pursuit of profits and the pursuit of growth are not identical goals, they are in many respects interrelated. *Economies of scale* permit larger producers to undercut smaller producers and thereby increase their market share, and this process in itself gives the larger producer more power and flexibility in dealing with consumers. In sectors requiring very large capital investment per unit of output, such as the basic resource processing industries (metal refining, petrochemicals, and pulp and paper), there are often added *technological* arguments for large-scale production. *Economies of scope* can be achieved by large multiproduct or multifunction firms which benefit from integrating a wide range of activities internally and reducing their transaction costs. As a firm's competitiveness becomes increasingly dependent on technological sophistication, requiring substantial investment in *research and development* (R&D) over a sustained period, there are further advantages which accrue to large producers who can recoup these outlays from high-volume sales. In many consumer goods sectors characterized by 'mature' products, such as detergents, the volume of sales which a particular firm can achieve depends on heavy *advertizing* expenditures to maintain its 'brand image', which again favours large, established producers. One or more of these characteristics can represent a *barrier to entry* which makes it increasingly difficult, if not impossible, for new firms to enter a given industry which has become dominated by a few large firms.

However, the concentration of output in large firms reflects more than just the technical characteristics of the production process. It first became noticeable in the USA in the last quarter of the 19th century, during the economic slow-down of the second Kondratieff cycle. This prompted some of the larger firms, for whom business risks had increased in proportion to their scale of activity, to abandon competitive behaviour in favour of collusion to maintain prices and profits (Mandel 1968). For such *oligopolistic* behaviour (in which the market is effectively controlled or administered by a few large producers) to be effective, it is necessary to concentrate ownership of productive capacity in a small number of business units. This was achieved in the USA by the epic capitalist trust-makers (men such as Rockefeller and Vanderbilt). The pattern was repeated in Germany in the early 20th century, where the State actively promoted the formation of large industrial combines as a means of hastening that country's catch-up to Britain's industrialization. (British

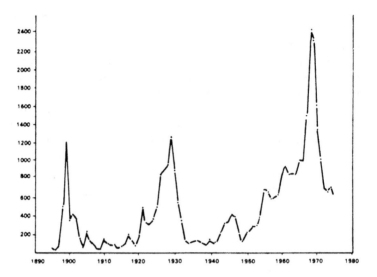

Figure 6.2   The long-term pattern of corporate acquisitions in the USA, 1895–1975.
Source: Edwards, G.A. 1977. Historical perspective on acquisition trends. *Foreign Investment Review* (Ottawa) 1(1); Chart 2.

capitalists remained much more strongly attached to their traditional independence as owners of small firms, and generally resisted moves to create larger manufacturing enterprises, which is one reason why British economic performance began increasingly to suffer in comparison to that of the USA and Germany.)

Initially, the most common mechanism of corporate growth was the *merger* (amalgamation) of individual companies, but the larger-than-average firms which resulted became increasingly active in the *acquisition* (takeover) of smaller enterprises. Acquisition activity in the major capitalist economies has exhibited dramatic variations over time, being concentrated in a few bursts. Edwards' (1977, p.13) analysis of US data (Fig. 6.2) argues that 'while a rapidly rising stock market seems to be a necessary component in any acquisition boom, some additional . . . combination of factors also seems to be required.' He identifies these factors in terms which strongly resemble Borchert's (1967) interpretation of the evolution of the US metropolitan system (see Figs. 4.4 & 5). The acquisition wave which peaked in 1899 followed the completion of nationwide railway, telegraph, and telephone networks which made it possible for companies to restructure themselves to exploit national, rather than just regional, markets, and at the same time to draw upon a greatly extended, national, capital market. The boom which peaked in 1929 marked the influence of the 'widespread ownership of automobiles (which broke down local monopolies) and radios (which permitted low cost, effective national advertising).' The acquisition boom which peaked in 1968 was associated particularly with the formation of industrial *conglomerates*, very large organizations involved in a wide variety of often unrelated product markets, the effective management of which was (theoretically) possible only because of postwar developments in managerial science and computer-based information systems. In practice, many conglomerates, such as Gulf and Western and ITT, did not prove to be outstandingly successful, when judged by the financial criteria which were the rationale for their creation.

Galbraith (1973) argued that the size and oligopolistic market power of large corporations, those in which are concentrated control over the majority of national productive assets in the Western industrialized states, gives them the ability substantially to shape, and not simply to react to, the economic and political environment in which they are set. He thus characterized them as 'the planning system' to distinguish them from the vastly greater number of smaller and weaker firms which make up the balance of national economies. Together, these latter comprise 'the market system', which exhibits to a much greater degree the characteristics of a competitive capitalist economy modelled by neoclassical theorists. There are, indeed, many significant differences between the dynamics of these two systems (more accurately, subsystems), but there is a danger that in concentrating on the undeniably influential behaviour of firms in 'the planning system' we overestimate their rôle in the economy.

Deconcentration tendencies in the late-20th century economy
Miller (1975) observed that in the USA, medium-sized firms (with assets in the range of $1–250 million in 1970) employed 40% of the civilian labour force and performed distinctive rôles within the national economy. Some were regular subcontractors to firms in the planning system; others had operations as influential within one or more geographical regions as those of large firms nationally. Moreover, although the majority of small businesses are in the service sector, fulfilling routine needs within a geographically restricted area, it is too easy to discount the significance of small manufacturing firms. This is particularly so in many high-technology fields, where products are characterized by embodied knowledge rather than resource-intensity, so that a firm can market a specialized product which is internationally competitive with relatively limited capital investment (although difficulties in securing adequate *working* capital

often result in innovative small firms being absorbed by a large corporation).

Small firms have attracted considerable interest in the 1980s among both academics (Mason & Harrison 1985) and public policy-makers in Western Europe and North America (*The Economist* 24 December 1983). Their recent record of job-creation has been very much stronger than that of the large firms who had previously been wooed by governments as stimulants to regional development (see Ch. 8). This differential performance has much to do with the greater concentration of small firms in the service sector, but it also reflects the changing structure of manufacturing in core industrial nations.

Large firms in the manufacturing sector, whether transnational or domestic, have prospered with the dominance of standardized, mass-produced goods and the efficient production technologies and routinized labour processes required to create them. These characteristics of *fordism* (a term derived from Henry Ford's innovation of assembly-line production) are now available most profitably in the NICs, where labour productivity is high but wages are low by core standards (Lipietz 1986). Whereas fordism in the core nations was accompanied by the growth of *domestic* mass consumption, as unionized workers fought for their share of the wealth created by higher productivity, fordism in the NICs has been geared overwhelmingly towards *export* markets in the core (where its output has been 'disruptive' of fordist production and employment). Industrial and civil unrest in Korea since the mid-1980s, leading to significant wage increases, suggests that a greater share of output in the most successful NICs will be directed towards an increasingly affluent domestic market, but the export thrust will be continued by 'upcoming' NICs, such as Malaysia.

The basis of manufacturing profitability in core nations has shifted towards systems of 'disintegrated' and 'flexible' production. The rigidities of fordist mass production (standardized products, narrowly specialized capital equipment, and task-specific labour) are a liability in a much more dynamic and competitive global economy and in consumer markets which are more volatile. Computerized production technologies, which can achieve high efficiencies in small-batch operations and can be reconfigured quickly, are the basis of these systems. Dominant firms contract out substantial amounts of production to a network of smaller specialist input suppliers (who themselves depend on a variety of such contracts), so as to achieve greater flexibility in changing their volume and type of output. More flexible labour contracts, in terms both of job assignment and security, characterize these production systems. Spatially, they encourage regional agglomeration, on account of the intensity of inter-firm collaboration, including 'just-in-time' delivery arrangements (Schoenberger 1988).

Historical development of corporate production systems

It is easy to underestimate the riskiness of early capitalist industry. The basic unit of business was the individual entrepreneur or the small partnership of friends or relatives. Financially and technically, most firms began on a very modest scale, with the accumulated savings of the proprietor(s) being spent on *relatively* inexpensive machines housed in makeshift premises employing perhaps ten workers. Certainly, there were firms which started out on a much more substantial scale, but it was well into the 19th century before modern forms of business organization were given legal standing. Not until the 1850s were public companies with limited liability accepted in Britain or did similar institutional arrangements become widespread in the USA (Viljoen 1974). On both sides of the Atlantic, the beginning of railway construction was the catalyst for creating corporations whose demand for fixed capital represented a quantum leap in the scale of business enterprise, and whose financing was instrumental in developing the rôle of the stock exchange (Reed 1969).

By the end of the 19th century the development of efficient and reliable transportation and communications systems in the industrialized core economies of North America and Western Europe made it possible for firms to structure their activities at a national scale. Leading manufacturing firms began to exhibit a distinctive corporate geography, within which increasingly specialized functions were allocated to different locations on the basis of their specific characteristics. Early-19th-century capitalists frequently carried out the full range of managerial tasks from an office in their mill or factory, but in the early 20th century the management of large industrial corporations, especially in the USA, passed increasingly into the hands of a team of assorted professionals. Improvements in the quality of products and the efficiency of production processes became more directly tied to scientific research and the development of new technologies. Nationwide marketing and product distribution called for new types of skill and organizational capability.

These components of the overall process of production had locational requirements which differed considerably from those of the manufacturing operation itself. As factories decentralized within metropolitan areas, managerial functions stayed within the city cores, where contact between interdependent decision-makers and access to sources of information and finance could be most efficiently maintained. Metropolitan skylines began to be dominated by head-office towers which housed the corporate administration, and made a symbolic statement of the firm's importance and permanence (Gottman 1966).

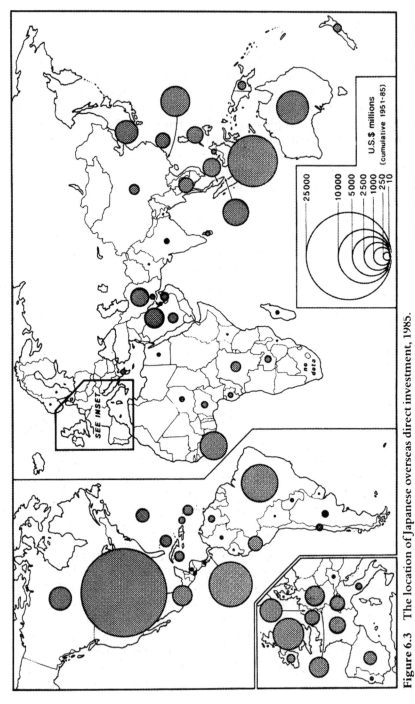

Figure 6.3 The location of Japanese overseas direct investment, 1985.

Source: Dicken, P. 1988. The changing geography of Japanese foreign direct investment in manufacturing industry: a global perspective, *Environment and Planning A* 20, Figure 4.

Research facilities, on the other hand, tended to be located on cheaper and environmentally more attractive sites on the metropolitan fringe. As for the goods themselves, the growing tendency for consumer items to become standardized and mass-produced for mass markets meant that regions distant from a manufacturer's 'home' increasingly justified the establishment of new production facilities to serve their local demand. Within multilocational firms, therefore, an internal core-periphery pattern of activity began to emerge. The 'home' (core) region retained the full range of corporate activities, including the strategic managerial and R&D functions, whereas the corporate periphery (which obviously included many large metropolitan centres) acquired only the capacity to produce standardized items.

Many of today's TNCs (which are multinational as well as multilocational) have a long history of foreign operations. Firms based in imperial European states (notably Britain, France, and the Netherlands), with establishments in one or more colonial territories, expanded and diversified their operations in the closing decades of the 19th century. In general, these corporations had interests in the primary and tertiary sectors of colonies of exploitation (minerals, tropical agricultural products, trade, and finance), being attracted to overseas territories by their specific resource endowments or the opportunities for monopolizing their external trade. British manufacturing firms did invest directly in production facilities abroad, primarily in colonies of settlement, but only on a limited scale. In contrast, foreign direct investment by leading US manufacturing corporations became increasingly prominent after 1870. The core economies of Western Europe represented a large market for products in which US manufacturers had either acquired a technological lead or else achieved notable economies of scale through rationalized and mechanized production systems. By setting up branch plants in Europe, US firms overcame the tariff barriers which would have limited their export sales. They also ensured that their operations were more sensitive to European market conditions in an era when trans-Atlantic communications were far slower and more limited than they are today.

The immediate aftermath of the Second World War witnessed a dramatic increase in the volume and geographical spread of investment in foreign establishments by US corporations. Not only was the USA the undisputed world leader in technological development, but the economies of European core states were in the throes of postwar reconstruction, and funds for investment overseas were extremely limited. Many medium-sized US firms which had not so far expanded their operations outside of North America opened branch plants in Western Europe, and also, on a much smaller scale, in Latin America, where European investment had previously been substantial. In Britain and parts of continental Europe, this influx of US direct investment coincided with the implementation of policies designed to attract manufacturing firms to peripheral regions suffering from above-average levels of unemployment. The foreign firms often displayed a greater willingness to locate in such regions than did domestic manufacturers, and they were the more welcome for that. So successful was the revival of the Western European economy in the 20 years following the Second World War, however, that from the late 1960s a substantial reverse flow of European investment into the USA became once more apparent, aided by the weakening of the dollar. Since the late 1970s, Japanese industry has become increasingly active as an investor in production facilities overseas (Dicken 1988). The rapidly increasing value of the yen and the growth of the non-tariff barriers against Japanese imports in other core markets has accelerated the spread of Japanese manufacturing foreign direct investment (FDI) beyond Asia (Fig. 6.3.).

Types of transnational corporation and the nature of their global orientation

The motives for establishing production establishments in a foreign country vary amongst TNCs. For firms in the primary sector, the dominant motive is undoubtedly to obtain access to exploitable resources. Manufacturing firms' primary interests are access to markets or to cheap labour. Manufacturing is the single most important sector of FDI in terms of capital stock, and the one in which by far the greatest proportion is concentrated in the industrialized world itself. From this it is clear that transnational corporations 'seek out and flourish most in foreign markets that most closely resemble the home markets for which they first developed their products and processes' (UN 1978, p.40). The relative importance of foreign investments in the primary and in the service sectors has switched since 1970. Nationalization of resource properties, especially in oil-producing nations, has reduced the capital stock of TNCs in those regions and inhibited further investment in this sector. At the same time, TNC activity in the service sector has grown rapidly, partly in direct response to the needs of manufacturing firms expanding abroad. For instance, US advertizing and accounting firms followed their domestic clients into Western Europe, and later into the NICs. Major banks have expanded and diversified their foreign operations very rapidly as host countries have loosened their restrictions on foreign participation in this sector. Loans to Third World governments and the financing of East–West trade have been added to the business of meeting the capital requirements of TNCs.

Primary sector

The majority of TNCs which exploit and market tropical agricultural commodities and non-agricultural natural resources are *vertically integrated*. The whole range of production activities, from extracting the resource, through processing and fabrication, to the delivery of a commodity to the final consumer is undertaken within a single corporation. The leading petroleum TNCs provide the most complete example. Although the characteristics of the global oil industry have changed substantially since the early 1970s, the 'majors' (Exxon, Shell, Texaco, Mobil, Standard of California, BP and Gulf) still play a leading rôle. The corporate geography of these firms is one in which control, planning, and research functions are concentrated in the home countries (the USA and the UK); crude oil production is scattered around the world in a set of geologically determined source regions, which vary from company to company but include establishments in both industrialized and underdeveloped states; refining and 'downstream' manufacturing activities (petrochemicals and goods made from them) are predominantly located in the principal markets of the industrialized core states; and marketing networks at the wholesale and retail level exist in almost every part of the non-socialist world, on a scale determined by the volume of local demand. The largest corporations in the non-fuel mineral sector have a similar structure, although one which lacks the final level of distribution networks and which displays very much weaker vertical integration at the manufacturing level. For instance, Inco, the leading Western nickel producer, operates mines, smelters, and refineries, as well as owning a number of nickel-using manufacturing firms. The strength of agribusiness firms such as Standard Fruit (bananas) lies in their mastery of the logistics of crop harvesting and international marketing.

Manufacturing sector

The global pattern of investment by TNCs in the manufacturing sector has become increasingly complex since the early 1970s. During the 1950s and 1960s, the overseas expansion of US-based firms dominated flows of FDI. Later in this period, Western European based corporations became increasingly active, followed in the 1970s by Japanese firms. Since 1980, there has also been a growing volume of FDI by TNCs based in less developed countries, mainly in neighbouring states (e.g. by Singapore-based firms in Malaya), but also in industrialized countries (e.g. by the Korean automobile manufacturer, Hyundai, in Canada). The motives for FDI have broadened over time. Until the early 1970s, the vast majority of overseas establishments were designed to extend a firm's sales beyond its domestic market. They were therefore concentrated in high-income regions (notably those of US TNCs in Western Europe) or in large Third World nations, such as Brazil, whose social élites created a sufficient volume of demand for core nation products. Since then, the search for dramatic reductions in labour costs for component assembly has prompted many TNCs to locate production plants in low-wage countries. An increase in protectionist measures by industrialized countries has also resulted in an increase in FDI, to circumvent import restrictions or to satisfy 'domestic content' regulations: the Nissan automobile plant in north-east England is a notable example (Dicken 1987).

The conceptual frameworks within which the geographical behaviour of TNCs in the manufacturing sector has been explained need to be evaluated in this context of a dynamic and increasingly complex global pattern (Dicken 1986). The emergence of new technologies and products and of organizational structures best suited to their profitable production, associated with Kondratieff cycles, suggests that interpretations of changing spatial distributions of manufacturing activity must be sensitive to their *conjunctural* (historically and geographically specific) origins. For example, the concept of the *product life cycle*, reviewed below, was developed when the domestic economy of US TNCs was by far the world's largest, richest, and most technologically advanced market for manufactured goods, as is suggested in Figure 6.4. In contrast, the concept of *the new international division of labour* (Fröbel et al. 1980) captures the dynamics of production that developed in the 1970s, when firms moved towards operating a global network of production establishments (often through subcontracting rather than FDI), located to exploit regional comparative advantages within a functional division of labour. Scott (1986) argues that we must recognize the conjunctural contingency of this pattern also: the diffusion of computer-assisted manufacturing and *just in time* component delivery systems may well create the basis for emergent industrial complexes in the leading industrialized nations.

The point of these comments is not to argue that conceptual frameworks cease to be enlightening as soon as new conjunctures develop, because changing conditions are not simultaneously nor as radically disruptive for different products, different firms, and different production locations. They should caution us, however, against the overgeneralized application of theories of manufacturing location.

The product life-cycle model assumes that manufactured goods 'undergo predictable changes in their production and marketing characteristics over time, . . . that the production process is characterised by economies of scale, that it changes over time, and that tastes differ in different countries' (Wells 1972, pp.5–6). It also assumes that there are geographical impediments to the flows of information which affect investment decisions, and that these favour the introduction of innovative products in the leading industrial economy, where demand is strongest, market information most complete, and capital and specialized skills are most readily available.

At the start of the product life-cycle, a number of considerations favour the location of manufacturing plants close to a firm's geographical core of central administrative and research establishments. The commercial introduction of a new product requires close co-operation between the R&D staff, production engineers, and senior marketing personnel. This phase is thus relatively labour-intensive and demands more direct managerial attention than do later phases. The relatively high unit cost of the product at this stage is not a major concern, because demand for its innovative qualities tends to be relatively insensitive to price. It is suggested in Figure 6.4 that initial manufacturing is confined to the corporation's home country, while the limited market abroad is supplied by exports.

The second phase of the product life-cycle is marked by the emergence of sufficient demand in other core states that local production for these markets begins. By this time the product has been refined and its production technology standardized, so that routine manufacturing operations can safely be conducted in locations remote from the corporate core. In the process, unit costs have been reduced, widening the effective market. The probability that other corporations either at 'home' or in other core states, will acquire the capability to produce a competitive product is a major incentive for establishing foreign branch plants, the output of which is designed to consolidate and expand the markets which have so far been supplied by exports from the firm's domestic establishments. This form of overseas investment enables the TNC to benefit from reduced transportation costs, better product servicing capabilities, avoidance of tariff barriers against imported goods and, most significantly, lower labour costs. (Note that the differential between US and Western European wage levels was quite marked until the late 1960s.)

The last three phases of the product life-cycle as depicted in Figure 6.4 are, above all, characterized by the growing 'maturity' of the product. It no longer embodies distinctive technological capabilities which command a high price: it is essentially a well tried item which still commands a mass (although eventually shrinking) market, but one in which price competition is severe. Its production technology is now thoroughly standardized and creates almost no demand for skilled labour. As a result, the bulk of corporate manufacturing capacity shifts steadily to increasingly low-wage countries in the Third World, from which the demand of global core markets is met. Meanwhile, the TNC's corporate core establishments are engaged in the early life-cycle stages of a new product.

Vernon (1966), who introduced the product life-cycle concept, assessed its continuing validity in a changed international environment (1979). First, he recognized that the TNC is no longer a pre-eminently US phenomenon. By 1979, the USA was the home country of fewer than half the world's largest industrial corporations, as measured by *Fortune* magazine's annual rankings of the top 100. European, Japanese, and even some Third-World TNCs are capturing global markets with innovative products which often reflect particular challenges faced in their home economies. Vernon noted, for instance, that the differing population/resource ratios of Europe and North America have fostered differing technologies (just as they did in the 19th century; see p.83) and that, historically, Japanese firms have developed 'products that conserved not only material and capital but also space' (1979, p.256). The US Conference Board (1981) saw a connection between rising energy costs in the USA and the opening of US production plants by such European firms as Michelin and Bosch, with their fuel-saving technologies of radial tyres and fuel-injection equipment respectively.

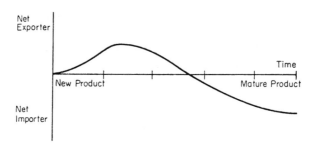

Phase I	Phase II	Phase III	Phase IV	Phase V
All production in U.S.	Production started in Europe	Europe exports to LDC's	Europe exports to U.S.	LDC's export to U.S.
U.S. exports to many countries	U.S. exports mostly to LDC's	U.S. exports to LDC's displaced		

Figure 6.4 A schematic presentation of the US trade position in the product life cycle, c. 1965.
Source: Wells, L.T. 1972. *The product life cycle and international trade.* Boston: Division of Research, Harvard Business School, 1972, p. 15. Reprinted by permission.

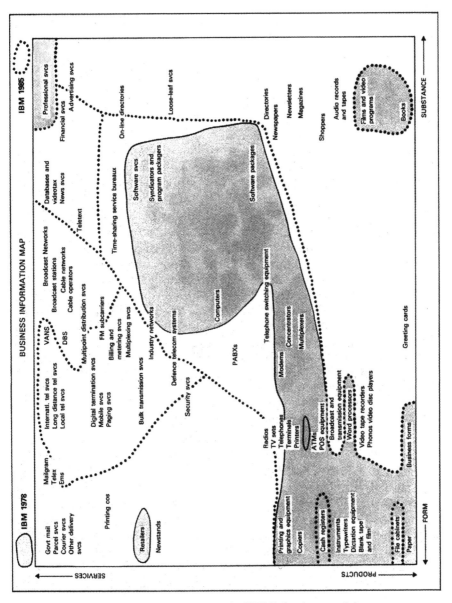

Figure 6.5 The changing 'corporate map' of IBM's business activity
Source: de Jonquieres, G. 1985. Activity map to stay ahead in the race. *Financial Times*,
28 November 1985. Based upon work of McLaughlin, J. F.,
Centre for Information Policy Research, Harvard University.

A second change has been the notable increase in the number of countries within which a typical TNC operates production establishments. One result of this is that the time lag between the introduction of a product in its home market and its diffusion into others has significantly shrunk. Another is the greater degree to which different components of a given product are produced at various places within the corporation's global system, for a variety of economic and political reasons (such as 'offset agreements' to soften the trade balance impact of major capital purchases). Vernon noted, however, that differences in production strategies and locational preferences during the product cycle will reflect different patterns of corporate behaviour and differences between products. Unfortunately, some of the applications of the product life-cycle concept by geographers have not been as discriminating, and fully justify Taylor's (1986) critique. Auty's (1984) study of 'the location of the global petrochemical industry after the second oil shock' illustrates the utility of the model if properly contextualized.

The concept of the new international division of labour equally needs to be used with care, as Thrift (1986) recognizes. Its essential characteristic is the growth of industrial employment in parts of the Third World (the NICs) which had very little history of export-oriented manufacturing prior to the late 1960s (see 'the pull of low-cost labour', Ch. 7). It has involved both the relocation of certain types of production (clothing, small domestic appliances, etc.) from core countries such as Britain, West Germany, and the USA to sites in NICs, as well as the location in these nations (especially in Asia) of new lines of production, especially those associated with microelectronics assembly (Sayer 1986).

Service sector
As the domestic economy of core nations becomes increasingly oriented towards services, firms in this sector are developing the global strategies of growth which characterize manufacturing TNCs. Services vary substantially in character, but many of them are 'location specific in terms of the close ties that are necessary between the providers and demanders': they therefore tend to require FDI in the host market rather than provision from a foreign 'home' (Stern 1984, p.182). The lifting of national restrictions on the participation of foreign firms in the service sector of host nations is one of the principal contentious issues in contemporary international trade negotiations. Recent moves towards the deregulation of financial services in developed nations have been stimulated by the rapid advances in computer and communications technology. The ability of institutions to carry out transactions on the other side of the world almost instantaneously has meant that financial centres whose regulations inhibit maximum flexibility of operation will

Table 6.1   National shares in world services exports, 1980.

Country	Shipment (%)	Other transport- ation and passenger services (%)	Travel and tourism (%)	Other private services (%)	Total[a] (%)
USA	7.9	15.9	12.2	9.8	11.7
Canada	1.6	0.9	3.5	2.4	2.2
Japan	16.0	7.5	0.8	4.7	6.1
UK	12.4	13.8	8.4	12.1	11.6
France	2.8	13.2	10.0	11.6	10.3
Germany	11.3	8.0	8.0	14.1	10.6
Italy	7.2	4.2	10.8	7.1	7.3
Netherlands	7.8	8.8	2.0	5.6	5.8
Belgium	5.2	2.8	2.2	7.7	4.7
Switzerland	0.9	n/a	5.0	3.6	2.6
Sweden	3.8	2.3	1.2	3.1	2.5
Norway	10.7	2.6	0.9	1.2	2.9
other countries[b]	12.4	20.0	35.0	17.0	21.7
totals[a]	46.5	81.2	82.4	109.9	320.0
% $US billions	100	100	100	100	100

[a] Based on totals for the 25 largest service exporters in 1980.
[b] Refers to 13 developing countries.
Source: Stern, R. M. 1985 Global dimensions and determinants of international trade and investment in services. In Trade and investment in services. Toronto: Ontario Economic Council; Table 5.

rapidly lose their rôle to those that attempt to keep abreast of change. The deregulation of financial services in the City of London in 1986 was prompted by precisely this concern.

The significance of British-based TNCs in sectors such as insurance and banking is a legacy of the global span and commercial orientation of the Empire. In 1982, British insurance companies had 675 offices abroad (more than those of any other nation) in 47 countries (Howells & Green 1986). During the 1960s, US-based institutions grew to dominate the banking sector, reflecting the economic hegemony of the USA and the central rôle of the dollar in the capitalist world-economy. By the mid-1980s, however, with the emergence of Japan as the world's leading creditor nation, Japanese banks emerged as the largest and an increasingly internationally oriented group. The USA holds sway in sectors pioneered by domestic firms, such as fast-food franchising, and in tourism (Table 6.1). Competition between TNCs based in a number of nations, both developed and NICs, is strong in the international engineering and contracting industries, which enjoyed booming markets in the Middle East and other oil-exporting nations in the wake of the oil price rises of the 1970s.

Two characteristics of the service sector make it difficult to assess its true size and significance in the international economy. One is the increasing fuzziness of the boundary between goods and services in many fields. The way in which IBM's central business activities flow across the boundary between hardware items, such as production of typewriters, and software or service items, such as computer programs and database-management services, is illustrated in Figure 6.5. The second is the varying degree to which large 'manufacturing' TNCs externalize their requirements for services as opposed to providing them 'in-house'. The more production becomes amenable to robotic or other forms of computerized execution, the more likely it is that corporate information and manufacturing control networks will be retained within the firm rather than contracted out to specialist agencies.

Conclusion

The ultimate test of a firm's success within the capitalist world-economy is that it generates adequate profits for its shareholders. The population of small firms is in constant flux, as some grow, many fail, and new ones continue to be established. Larger firms have a greater capacity to weather fluctuations in the business environment, but many of them have found it difficult to adjust to the more intense global competition of the contemporary era. In the 1970s, governments in core states frequently came to the assistance of large corporations in financial difficulty, to safeguard employment or particular technological expertise; but there is much less enthusiasm for State intervention today. The late 1980s saw two particular developments, both of them emphasizing the fully global nature of the modern corporation. One was strategic alliances (Cooke 1988) between TNCs of different 'nationality' to achieve specific goals (e.g. of Toshiba and Thorn–EMI to manufacture compact disks). The other was a new round of very large takeovers to create globally powerful firms, in sectors such as publishing and information systems (e.g. the Maxwell and Murdoch 'empires'), confectionery (e.g. Nestlé 'swallowing' Rowntree), and beer (e.g. the global spread of Elders-IXL's Fosters).

7 Transnational corporations in a world of sovereign states

Power, leverage, and geopolitics

When, in the 1960s, the global hegemony of the USA appeared to be still secure, its technology triumphant (e.g. in space), and its corporations conquerors of markets around the world, it became popular for the executives of US TNCs to write the obituary of the nation–state. Among many quotable comments recorded by Barnet & Muller (1974, pp. 18–21) are those lamenting the parochialism of nations, their obtuse and reactionary commitment to independence, and the obsolescence of the political divisions of the world which they perpetuate. In contrast, the commitment of the TNC to bringing about a 'global optimization of resources', and its functioning as an 'instrument of world development ... and the only force for peace', was claimed to ensure that the rôle into which it was 'being pushed by the imperatives of [its] own technology' promised nothing less than prosperity and happiness for all. The earlier corporate ideology that 'what is good for General Motors is good for the United States' was being recast on a global scale. But as firms based in, and managed from, the USA became increasingly involved in the domestic economies of other states (both in the European core and the Third World periphery) as a result of their investments in foreign production facilities, governments of host countries were awakened to the need to come to grips with the impact of TNCs. The reality of potentially conflicting goals and of differential power to achieve them means that the relationships between sovereign states and powerful corporations play an important part in shaping the geography of the contemporary global economy.

The modern TNC possesses, to a historically unprecedented degree, that *structural* freedom of action which accrues to economic enterprises in a capitalist world-economy characterized by fragmented political juris-dictions. Advances in transportation and communications technology, and in the power and scope of information-processing systems, have provided the material preconditions for a relatively small number of firms to attain dominant positions, first on a national and then on an international scale. In the following sections we analyse the nature of the corporate interests which have determined their involvement with states, focusing particularly on the search for resources, for markets, for production cost advantages, and for trading profits. It will be apparent that the indisputable economic power of leading TNCs is further consolidated by the pervasive and often intangible influence which they derive from it. On the other hand, there is evidence that individual states are far from completely powerless to channel the activities of TNCs along lines which increase the benefits accruing to their domestic economies. In general, core states have more leverage to influence corporate behaviour than do peripheral ones, but under certain conditions even peripheral nations can bring substantial pressure to bear on TNCs operating within their territorial limits. Certainly, those corporate executives who wrote the obituary of the nation–state did so prematurely.

The relationship between government and private business in mixed-capitalist economies is a complex one. Government action is doubly constrained: by popular political pressures to protect individuals or communities from corporate actions and their externalities; and by economic pressures which require the State to act in the interests of maintaining a healthy business sector. The institutional responses to this tension differ between countries. Despite the widespread enthusiasm for 'privatization', which saw the State shrink during the 1980s, the extent of direct government ownership of industry remains greater in Western Europe and Canada than in the USA, where, 'to an exceptional degree, regulated, subsidized, and risk-underwritten private enterprise has been preferred to public enterprise' (Herman 1981, p.167).

Elaborating on US experience, Herman notes:

> . . . a variety of 'problems' are spun off by business in its process of growth – such as structural unemployment, waste disposal sites – which it does not internalize but compels government and the larger society to deal with as best they can. Thus, on the one hand, business has contributed greatly to the rise of big government by its own externalization of social costs and ad hoc mobilization of government – most often to restrain, or offset by subsidy, competition from domestic or foreign producers. On the other hand, business has felt harried and threatened by increasing government intervention, even when of its own creation . . . . Business is further troubled over the growth of governmental social welfare intervention – admitted to be necessary

in a complex social order but regularly criticized for overgenerosity and mismanagement (Herman 1981, pp.163–4).

Interchange of personnel between the US public and private sectors, especially within the 'military–industrial complex' (see p.262), has fostered a community of interest in many areas of policy. Even where the State appears to act as public watchdog, traditional forms of regulating industry (e.g. the Interstate Commerce Commission) have tended to evolve into a set of relationships within which business lives quite comfortably. During the 1970s, a new mode of regulation emerged which industry found more disturbing. Typified by the National Environmental Protection Act, legislation marked by 'its more detailed, multi-industry intrusions into areas of long-standing managerial discretion and [by] its cost- and liability-enlarging potential' (Herman 1981, p.185) changed the context for much corporate activity.

The privatization of large public-sector enterprises, such as British Airways, British Telecom, Conrail (USA), and de Haviland (Canada) leaves governments still major players in, and not merely referees of, their domestic economies. (This is overwhelmingly the case in the Third World; see Ch. 11.) Borchert (1978, p.229) points out, for instance, that the US Department of Defense alone 'appeared in 1971 as a conglomerate with assets of 133 billion dollars, approximately seven times the total assets of General Motors at that time.' In the UK, public undertakings such as the Post Office and British Rail rank in employment terms above the largest private business corporations. In Canada, government-owned utilities (Hydro-Quebec and Ontario Hydro) lead the assets ranking of leading industrial corporations. If one adds public-sector employees and expenditures in government administration at all levels, in education, and in health-care delivery, the pervasive presence of the State in the industrialized capitalist nations becomes very apparent.

The economic geography of these states thus reflects the complex interplay of the pursuit of profit, 'the public purpose' (Galbraith 1973), and individual wellbeing. The concentration of economic power in the hands of a minority of large business corporations and the spending departments of national governments has, at one level of analysis, led to a parallel spatial concentration (often more apparent in the public sector than in the private) in the location of influential decision-making establishments (Borchert 1978, Pred 1977). At the same time, in the USA more so than in Europe, rapid growth in the national economy has been accompanied by relative decentralization of economic activity away from traditional core regions. Government spending on goods and services, and the interpersonal and interregional channelling of income flows represented by public taxation and transfer-payments, have had complex geographical repercussions, sometimes reinforcing, sometimes counteracting, the evolving spatial patterns of economic activity generated by corporate and household decision-making (see Ch. 8).

The control of natural resources

During the first century and a half of the Industrial Revolution, the exploitation of natural resources generally took place with little direct state involvement. Even where the State was the legal owner of forest and mineral resources, governments generally contented themselves with collecting rents or royalties and left it to individual (and usually small) firms in the private sector to decide where, when, and how resource exploitation should occur. By the end of the 19th century, however, a number of more powerful firms began to emerge in the mining and petroleum industries, and the governments of core states began to take a greater interest in these industries because of their strategic significance in an era of mechanized warfare. The growth of a few dominant firms in these sectors at this time reflected the increasing importance of applied science, resulting in proprietory (firm-specific) technologies, in solving problems of converting natural resources into specific end-products. The act of harnessing the vast and diversified domestic resource base of the USA in support of that country's vigorous economic growth gave US corporations the financial strength and practical expertise to assume world leadership in most aspects of the mining and petroleum industries. In the years following the First World War, the growing resource needs of an expanding global economy encouraged the development of sources of supply outside of the core economies of the USA and Europe, in territories which were either formal colonies of imperial powers or which were functional dependencies of them. With their monopoly of capital and know-how, US (and a few European) resource companies were in a position to undertake resource extraction very much on their own terms in the overseas territories where they started production. No better illustration of these relationships can be provided than that of the international oil industry (Odell 1981).

When the transnational oil companies began to develop the reserves of the Middle East in the 1920s they encountered weak state authorities who granted them *concessions*, giving them exclusive rights over the development of the resource base and making them:

the sole arbiter[s] of the volume and nature of investment in the host country, the choice of areas for exploitation, the determination of exploration plans, the development of oilfields, the production levels, the size of the necessary production facilities, exportation and transportation capacities, etc. In practical terms this deprived the state of the right to interfere in any of these vital matters and limited its role merely to that of collecting taxes (Al-Chalabi 1980, p.9).

There was no obligation on the part of the oil companies to foster indigenous expertise in the resource sector, nor to contribute more generally to the economic and social development of the host country. Moreover, the world market for oil was tightly controlled by the major TNCs, by means of their overt collaboration in crude oil production. Instead of holding concessions from a host government directly, they usually established a consortium charged with producing oil on a break-even (non-profit) basis, so minimizing the state's income. The volume of production which this consortium delivered to its sponsoring corporations was determined by their individual assessments of what their (competing) worldwide distribution networks could profitably market. By these means, the danger of overproduction, and hence of pressure to reduce prices and profit levels, was largely eliminated.

The geography and institutional structure of the oil industry changed decisively in the 1950s. Additional oil companies appeared on the international scene: either 'independent' US firms, established in their domestic market and now encouraged by tax incentives to expand overseas, or else newly established European firms, such as the state-owned Italian enterprise, ENI. These companies sought out new sources of oil, and accepted alternatives to the concession arrangement in their dealings with host governments, who thereby gained a larger stake in the management and revenues of oil production than had traditionally been the case. The development of the North African oilfields in Algeria and Libya was a particularly significant outcome of this new international oil regime, for it coincided with the outbreak of hostilities which closed the Suez Canal and profoundly altered the economics of transporting oil to Western Europe from the Arabian Gulf (Fig. 7.1). In order to gain outlets for their oil in the principal markets of North America and Western Europe, the corporate newcomers undercut the controlled prices of the oil 'majors', which they were able to do and still make an attractive return on their investments. They thus initiated the steady decline in the real price of 'world' oil throughout the 1960s which reflected the disruption of the institutional monopoly (strictly speaking, oligopoly) of the 'majors', and the erosion of the geographical monopoly of the traditional oil-exporting states.

Attempts by host governments to assert greater managerial control over their petroleum resources and to take a larger share of the revenues derived from their exploitation met with varying degrees of success prior

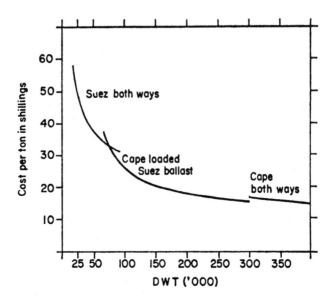

Figure 7.1   The unit cost of shipping oil from the Persian Gulf to Western Europe, by size of ship and route (mid-1960s data).
Source: Couper, A.D. 1972 *The geography of sea transport*, London: Hutchinson; Figure 6.

to 1972. But the changing institutional structure of the international oil industry during the 1960s enabled state authorities (often newly independent) to increase their control, and joint ventures between national oil companies and foreign TNCs became more common. The marketing monopoly of the oil 'majors' had been severely eroded by the 'independents', and national oil companies readily found distributors in the industrialized nations. Meanwhile, the demand for petroleum in the Western core nations grew rapidly, as a result of sustained economic growth and the displacement of coal. By the early 1970s, not only was oil consumption in Japan and Western Europe almost totally dependent on imported petroleum, but the USA, although still the world's largest producer, was importing almost 30% of its oil needs. This degree of dependence by the industrialized capitalist nations on petroleum imports (the socialist block remained self-sufficient), of which the majority came from Arab nations, gave the latter the economic leverage to transform the international oil regime and extract very high rents (see Ch. 3) during the decade following 1973. The host governments of OPEC nations were able to transform the oil companies, which had previously appropriated most of the resource (economic) rents themselves, into something much more like tax collectors on behalf of the state. With their production assets nationalized, the international oil companies became, in effect, marketing agents of various national (e.g. Iraqi, Kuwaiti) oil companies.

The predictable result of the OPEC high-price regime was to stimulate exploration for oil elsewhere and render many previously marginal oilfields profitable. North Sea, Mexican, and other reserves were developed, partly by TNCs, partly by public-sector companies in the home country. Alternative fuels, especially coal, became more competitive, and many of the oil 'majors' bought firms producing them. But slower economic growth, the spread of energy-conserving technologies, and the increased availability of non-OPEC oil (together with the vastly increased State expenditures of OPEC nations which has deterred them from significantly limiting production) undermined OPEC's monopoly position, reflected in the halving of 'world' oil prices in 1985. The power of TNCs in the oil industry has been permanently curtailed (as compared to the 1950s), and the firms are subject to greater State monitoring both in host countries and home markets. The power of mining TNCs, particularly in Third World nations, has been similarly reduced by more effective government regulation of the industry (Mikesell 1979).

Access to markets

Despite the recent changes in the global strategy of transnational manufacturing corporations noted in Chapter 6, markets remain the most important single locational attraction for foreign investment in this sector. Until the 1960s, foreign-owned branch plants were predominantly subsidiaries of US-based TNCs located in high-income countries such as Canada, the UK, and West Germany; but since then a growing number of branch plants owned by European and, more recently, Japanese manufacturing firms have been established in foreign industrialized nations (Hamilton 1976). The corporate motivation behind this form of expansion is typically to generate new sales, or to protect existing export sales against competition from manufacturers in the host country who may enjoy some measure of tariff protection. For many years, public authorities paid little attention to the inflow of FDI in manufacturing. Even when the number of US branch plants opening in Canada and Western Europe (especially the UK) increased dramatically after 1945, government awareness of the implications for national economic growth and its geographical distribution developed only gradually. Today, the issues raised by TNC investment are much better understood, partly as a result of the careful analysis by economic geographers documented below.

The impact of foreign-owned firms on a host economy is obviously related to the proportion of national manufacturing capacity which their operations represent. Hence, despite the concern voiced in the USA in the mid-1970s about the influx of foreign capital, especially when it was used to acquire existing US-owned manufacturing firms, McConnell (1980, pp.259–60) reported that 'foreign-owned companies account for no more than about six percent of the total output of any two-digit [SIC classification] manufacturing industry.' On the other hand, the composition of foreign investment rarely matches that of the domestic manufacturing sector as a whole. McConnell points out that most foreign firms investing in the USA are relatively large ones in oligopolistic, and especially high-technology, industries. The disproportionate significance of US-owned firms within the UK economy is indicated by the fact that in 1968 they accounted 'for only 1.4 percent of all private manufacturing establishments. Yet this minuscule proportion employed one out of every fourteen workers in United Kingdom manufacturing and produced 10.4 percent of total net output' (Dicken & Lloyd 1976, p.688). The situation in Canada, where in 1963 US-owned manufacturing firms embodied 46% of the sector's capital stock and provided 23% of its employment (Ray 1971b), faces policy-makers with issues which are not only quantitatively, but qualitatively, different from those raised by branch-plant investment in countries with a large and diversified domestic manufacturing industry.

Studies of the establishment of foreign branch plants, of their location within host countries, and of their linkages with suppliers have revealed various forms of geographical ordering. Collins (1966) found that 54% of European-owned branch plants opened in Toronto between 1945 and 1961 had started life as sales offices, and only 24% had begun as full manufacturing operations. Often, an initial sales unit had subsequently expanded to include warehousing functions, and then had become responsible for increasing degrees of product assembly (largely dependent on parts imported from the foreign parent), before completing the transition to a full manufacturing establishment. This gradual penetration of the Canadian market by European firms contrasts with the more immediate commitment to Canadian production by US firms. Collins cites a 1932 study which found that only 18% of US branch plants had grown from sales offices, and fully two-thirds had originated as manufacturing units. The difference in corporate behaviour can be viewed as a reflection of the lower level of uncertainty experienced by US firms entering Canada, as compared to European firms, stemming from their greater proximity to, and cultural similarity with, the host country. Such differences are narrowing, however, as European TNCs grow in size and experience, and as their internal style of management and their external business environment become increasingly similar to those of US-based firms. Linge & Hamilton (1981) quote Stopford (1976, p.26) to the effect that 'the development of important stakes in high-income markets during contemporary conditions requires more than an evolutionary transition from exporting to local production. Locally entrenched competition does not allow the new entrant time to evolve slowly.'

The spatial distribution of branch plants within host nations has been shown to differ from that of domestically owned manufacturing establishments, even when the comparison concentrates, as appropriately it should, on the distribution of the larger domestic firms. Disputes about whether the difference is significant and, if it is, whether it is beneficial or detrimental to the host economy, are frequently rendered inconclusive by the problems of obtaining suitable data. This is particularly true of analyses of manufacturing investment by foreign TNCs in the UK, where postwar governments have actively sought to encourage the location of new investment in regions of high unemployment. Watts (1979) compares UK employment data for the 13 largest British private-sector corporations and for foreign-owned corporations. Although US-owned establishments, accounting for 75% of the foreign-controlled employment, are over-represented in the prosperous South East region compared to large domestic firms, they are at the same time responsible for a disproportionately high level of employment in the most disadvantaged peripheral regions (Wales, Scotland, and Northern Ireland). The data suggest, therefore, that the spatial distribution of branch plants of US-based TNCs has contributed to the British government's goal of reducing regional disparities. The employment statistics also indicate that other foreign (non-US) TNCs have distinctly different location patterns within the UK. Branches of EEC-based firms show a notable preference for the South East and East Anglia, the two British regions closest to their home countries. Japanese investment has been proportionately most noticeable in South Wales.

The location of US-owned branch plants within Canada exhibits marked geographical ordering, with profound consequences for the distribution of economic activity across the country (Fig. 7.2). Seventy per cent of US branch plants operating in Canada in 1962 were concentrated in Toronto and southwestern Ontario (Ray 1971a, b). This region, part of which has traditionally served as a direct transportation corridor between Buffalo and Detroit, is a contiguous extension of the US manufacturing belt. Ray identified three locational factors which account for this highly concentrated distribution. First, Toronto is the point of highest *market potential* in Canada, providing US manufacturers with their easiest access to the bulk of the Canadian market. The number of branch plants located there reflects a gravity-type interaction process, being proportional to the size (manufacturing employment) of the US metropolitan centre containing a plant's head office and inversely proportional to the distance of that centre from Toronto. Thus firms based in New York, Chicago, and Boston, cities approximately equidistant from Toronto, had established 210, 147, and 14 branch plants in Ontario respectively. On the other hand, Los Angeles, although a much larger centre of manufacturing than Buffalo, was corporate home to only one more Ontario subsidiary than the border city (35, as against 34).

Secondly, Toronto acts as an *intervening opportunity*, whereby US firms rarely establish a Canadian branch plant which is more remote from their corporate home than is Toronto. As a result, areas to the north and east of Toronto (including the major industrial centre of Montreal) lie in what Ray terms an *economic shadow*. Thirdly, the location of US branch plants within Canada exhibits differential *sectoral penetration*. Corporations based in Detroit or Buffalo are likely to find that the ease of managing a factory which is only a few miles away across the border more than compensates for its not being somewhat closer to the centre of the Canadian market. Conversely, the more distant the US head office location is from the border, the more likely it is that the Canadian subsidiary will be situated close to Toronto, where access to markets and to international air services is maximized.

Host governments in industrialized countries are frequently more concerned about the integration of foreign-owned branch plants into the domestic economy than they are about their location. Multinational corporations which make substantial domestic sales involving a significant import content, notably in high-technology sectors, can create adverse balance-of-payments effects. Insofar as branch plants rely on flows of 'in-house' goods and services provided by the corporate parent, they create fewer demands for the output of local firms (Britton 1976) and generate employment characterized by a *truncated* occupational profile, one which lacks senior managerial positions or R & D functions (Hayter 1982).

Nations such as the UK and France are in a much stronger position to exert pressure on TNCs to organize their business in ways which promote national interests than are nations in the Third World which face similar problems. The allocation of public sector spending is itself a powerful instrument to encourage desired patterns of behaviour. The organization of IBM in Western Europe represents one response to demands that it act as a 'good corporate citizen', albeit at the loss of some efficiency. Production plants are 'so located that value-added in a given country is in fair proportion to IBM sales there .... Some effort is made also to ensure that bought-in parts come from countries where there is no direct manufacturing' (*The Economist* 29 October 1977, p.92). In Canada, much US investment used to be in *miniature replica* branch plants, manufacturing a full line of corporate products in small numbers (hence at high unit cost), adequate to serve only

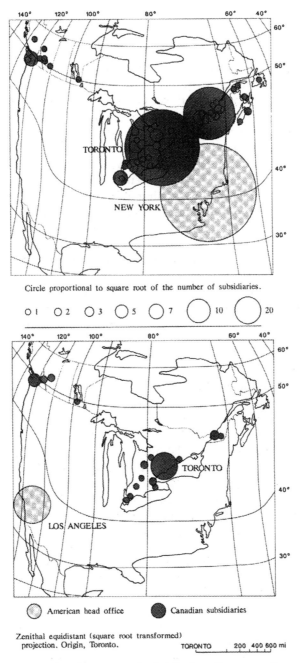

Circle proportional to square root of the number of subsidiaries.

◯ 1 ◯ 2 ◯ 3 ◯ 5 ◯ 7 ◯ 10 ◯ 20

American head office ◯ Canadian subsidiaries ●

Zenithal equidistant (square root transformed)
projection. Origin, Toronto.

TORONTO 200 400 600 mi

Figure 7.2 The Canadian locations of subsidiaries of US manufacturing firms based in Los Angeles and New York.
Source: Ray, D.M. 1971a. *Dimensions of Canadian Regionalism*, Geographical Paper no. 49, Ottawa: Department of Energy, Mines and Resources, Figures 30 & 31.

the needs of the domestic market. This pattern of TNC operations, a rational response to traditional Canadian tariff policy, leads to a stagnant and truncated manufacturing sector which is uncompetitive internationally and increasingly vulnerable to import competition at home (Britton & Gilmour 1978). One strategy which some TNCs have adopted that satisfies both corporate objectives and the desires of the Canadian government to see a more competitive and export-oriented manufacturing industry is the allocation of *world product mandates* to Canadian subsidiaries (Science Council of Canada 1980). Being given exclusive responsibility by the parent company for the development, production, and global marketing of a specific product or line of products allows the subsidiary to engage in the full range of managerial and R&D functions and to specialize in high-volume production of items which are technologically and economically competitive. Canadian demand for corporate products which are not produced domestically is met from similarly specialized production units in other parts of the world. Experience in Canada indicates that government assistance at a critical stage can be influential in securing a corporate commitment to reorganize its production on a world product mandate basis.

The pull of low-cost labour

Although the less developed nations of the Third World have become increasingly significant markets for TNCs based in the core states, it is low-cost manufacturing labour that has stimulated the dramatic expansion of industrial capacity in the NICs since the early 1970s. European, Japanese, and US corporations have established global production systems which are regionally and functionally differentiated on the basis of location-specific comparative advantage. Their direct investment, together with the forging of extensive subcontracting links between TNCs and NIC-based firms, has promoted export-oriented manufacturing development in an expanding range of product lines. The most spectacular growth has taken place in a set of about ten nations, led by the Asian group of Taiwan, South Korea, Singapore, and Hong Kong, together with Brazil and Mexico. The impact of TNC investment on host economies in the Third World is considered in detail in Chapter 11. Here, we focus on the corporate considerations which have prompted the expansion of manufacturing in low-wage countries.

Fröbel *et al.* (1980, p.13) identify three decisive developments which have provided TNCs with the incentive and the means to create production systems incorporating establishments located in the global periphery. The first is the emergence of 'a practically inexhaustible reservoir of disposable labour' in the less developed countries. This workforce is very cheap in comparison to that available in core countries, partly because individuals can so 'easily be replaced', and partly because it is unprotected by the legislation through which workers in Europe and North America have, during the past 150 years, achieved substantial security from exploitative working conditions and the vicissitudes of the business cycle. (Employers' contributions to unemployment, health, and accident insurance, representing part of the *social wage* of workers in the Western industrial nations, add significantly to nominal wage rates. Health insurance alone, for instance, was costing the Ford Motor Co. in 1979 'more per car produced than the steel that goes into each automobile' [quoted in the *Financial Post*, Toronto, 28 April 1979].) Given the universality of cheap labour, the specific identity of the NICs cannot be explained by this factor alone. Political, social, and economic conditions within these states are crucial to their definition as an attractive production environment (see Ch. 11).

The second development is that 'the division and subdivision of the production process is now so advanced that most of these fragmented operations can be carried out with minimal levels of skill easily learnt within a very short time.' As TNCs tend to use very similar technologies in their domestic and (export-oriented) foreign establishments, it follows that the savings which they can achieve through lower wage costs in less developed countries may not be significantly eroded by a lower level of labour productivity than they experience in their home-country operations. Although savings are most noticeable in relatively labour-intensive industries, such as clothing and microelectronic assembly, they are still substantial in more capital-intensive industries. Calcagno & Knakal (1981) cite statistics which indicate that although comparable steel industry products require 4.5 times more labour inputs in Brazil than they do in the USA, Brazilian labour costs are only one-tenth of the US equivalent.

The third development is the revolution which has taken place since the late 1960s in intercontinental freight transportation and communications technologies. Containerized shipping, wide-body airfreight jets, and satellite telecommunications have enabled TNCs to achieve the efficient integration of production units scattered across the globe, even in a time of rising fuel costs. It is not only light, high-value-added items, such as electronic components and the products made from them, which can withstand the costs of being shipped halfway around the world. Fröbel *et al.* (1980) document the extensive use made of air transport by West German clothing firms importing standard items such as shirts from subcontractors in the Asian NICs.

Clearly, there are costs as well as benefits to Third World countries which attract export-oriented manufacturing investment by TNCs.

Governments of NICs have welcomed foreign corporations on the basis of the export earnings and the employment which they generate. The degree to which the benefits of economic growth have reached the population at large varies considerably (see Ch. 11), with Taiwan, for instance, having a much better record than Brazil in this regard (Chenery 1980). In many cases, the TNCs' low labour costs are achieved only at the expense of industrial working conditions which approximate the excesses of the early Industrial Revolution in Great Britain (see Ch. 4). Fröbel *et al.* (1980) cite Hong Kong census statistics which recorded nearly 25 000 children between the ages of 10 and 14 in full-time employment in 1971, primarily girls in the textile and garment industries. In these sectors, the average hours of work per week of all employees was 48, compared to less than 40 in the USA and Japan. (The dilemma of the Hong Kong authorities, faced with a massive influx of population from China, needs to be appreciated, however, and its achievements in the areas of job creation, housing, education, and health services recognized.)

The leverage of Thirld World governments over TNCs which are primarily interested in a cheap-labour location is limited by the fact that so many states can potentially provide that requirement. Just as rising wage levels in Japan in the 1970s helped to accelerate Japanese manufacturing investment in the Asian NICs, so in the 1980s low-skill jobs are leaving Taiwan and Singapore for cheaper workforces in Malaya, the Philippines, and Bangladesh. Such mobility is facilitated by the very low fixed capital requirements of the manufacturing processes involved (Calcagno & Knakal 1980). Less developed nations which have gone all out to attract branch plants or subcontracting establishments to *export production zones* (EPZs, or 'free trade zones'), have begun to recognize some of the limitations of this strategy. An EPZ is an area in which the state provides the basic infrastructure required by (primarily) labour-intensive manufacturing units and allows the firms which locate there to enjoy 'duty-free imports of raw materials, components, and machinery required directly in the manufacturing process, and minimal customs formalities for the export of finished or semi-finished products' (Singh & Choo 1981, p.494). The absence of 'red tape' is as important to many TNCs as the low cost of labour. Factories in EPZs tend, however, to develop few linkages, either between themselves, because their operations are essentially similar, or with the rest of the host economy. These 'industrial mono-structures' (Fröbel *et al.* 1980, p.328) contribute only marginally, therefore, to the development of an indigenous manufacturing capability.

During the 1970s, an increasing number of Western European TNCs began to tap a closer source of low-cost labour than Third-World NICs, the workforce of the industrialized socialist nations of Eastern Europe.

These host countries have more power than do the majority of Third World states to control the terms governing foreign investment or trade. Wholly owned branch plants have not generally been permitted, so joint ventures and subcontracting arrangements are the principal channels of corporate involvement. Although the Eastern European regimes might be expected to ensure that TNC activities conform more fully with national development goals, Fröbel *et al.* suggest that the eagerness to promote export-oriented manufacturing has led some socialist states to surrender more of their freedom of resource allocation than is in their best interests. The nature of the exported goods, which have to satisfy Western standards of quality, calls for preferential treatment being given to the production establishments involved, at the expense of investment in the domestic manufacturing economy. This distortion of priorities has to be weighed against the desired consequences of TNC involvement (hard currency earnings and access to Western technology and know-how) in evaluating the host country's benefits (see Ch. 9).

Flexibility of capital movements

Transnational corporations possess the ability to capitalize on structural characteristics of the capitalist world-economy in ways that allow them to circumvent the authority and controls of sovereign states. Freedom to channel financial resources without government regulation, and to protect revenues from taxation, is a particularly strategic feature of TNC operations. Here we focus attention on two aspects of that freedom, the phenomenon of *transfer pricing*, and the creation of the *Eurocurrency market*.

Transfer pricing

Conventional theories of international trade, such as that of Hecksher–Ohlin (see Ch. 3), rest on the assumption that goods moving in international trade do so at prices and in volumes which reflect competitive market conditions. However, the growth of TNCs has meant that a significant component of global trade involves *intra-corporate* transactions, the volumes of which 'may well be determined on considerations of comparative cost, at least from the [TNC's] point of view, but the values stated may be quite different from those in open-market conditions' (Lall 1980, p.128). Moreover, for many items, especially high-technology goods embodying proprietary corporate innovations, it is very difficult to reach a fair 'market' evaluation. Under these conditions, TNCs may assign transfer prices for international transactions between production units which allow them to take their profits in whichever country (at the corresponding stage of the production process) it is most advantageous to do so. Global corporate profitability can be maximized in this way, and specific government restrictions (e.g. on the remittance of profits from a foreign branch plant to the home country) can be evaded.

Third World nations are most vulnerable to being exploited by transfer pricing, primarily because they lack the resources and expertise to monitor intra-corporate trade and so detect flagrant examples of undervalued exports or overvalued imports. The authorities in the industrialized core states (which are also, of course, the home countries of most TNCs) are better equipped to protect their national interests, and in the case of Canada and the USA have gone to the stage of undertaking joint tax audits of selected corporations. The international pharmaceutical industry has attracted particular attention, and Lall (1980, p.125) reports a detailed study of its transfer pricing in Colombia. This found 'that the weighted average of [corporate] overpricing ranged from 33 percent to over 300 percent for the imports investigated in the pharmaceutical sector, and from 24 percent to 81 percent in the other sectors.' For the majority of firms investigated, imputed overpricing profits exceeded declared profits.

The Eurocurrency market
Starting in the 1960s, and accelerating through the 1970s, there has emerged an 'unregulated stateless pool of money' which by the early 1980s was estimated to amount to $1000 billion (Muller 1980, p.24). The term 'Eurocurrency market' is now a misleading designation for what began as deposits of dollar earnings of US TNCs with European banks, but which has become a multicurrency fund handled by multinational banks through centres such as New York and Singapore, as well as the financial capitals of Europe (Thrift 1986). The significance of the Euromarket is that it provides large corporate borrowers with access to funds which are beyond the reach of a government's attempt to control credit or the domestic money supply, thereby making national macro-economic management that much more difficult.

During the 1970s, the volume of Eurocurrency funds was vastly expanded by the OPEC nations, which recycled their rapidly rising oil revenues by depositing them with major Western banks. These funds were then loaned to a variety of public and private sector borrowers to finance trade and investment. The recycling mechanism of the Eurocurrency market played a large part in helping the capitalist world-economy to adjust to the financial 'shocks' of oil price increases in the 1970s, but the benefits were distributed differentially. Demand for goods from TNCs based in the industrialized nations of the West was maintained at the cost of growing indebtedness among Eastern

European and non-oil-producing Third World nations. Moreover, this debt became much more closely tied to commercial interest rates than to preferential rates granted by individual states or international lending institutions such as the World Bank. The implications of the resultant 'debt crisis' for the borrowers are discussed in Chapter 12.

Prospects

Previewing the findings of Dunning & Stopford (1983), *The Economist* (19 February 1983, p.86) argued that 'Multinational companies are coming in from the cold. Governments in both rich and poor countries, anxious for new investment to promote jobs and exports, are giving multinationals a warmer welcome.' The adversarial relationship between states (especially in the Third World) and TNCs, which characterized much of the 1970s, has softened partly for the reason just cited and partly because governments have become more adept in their dealings with foreign investors. For instance, whereas TNCs are still in a position to play one state off against another in seeking investment incentives, the proliferation of corporations capable of operating at a global scale from a variety of home countries makes it easier for a potential host country to bargain with competing would-be entrants to its domestic market.

A further development tending to reduce tension between host governments and TNCs is a continuing sectoral shift in patterns of FDI towards manufacturing and services, at the expense of natural resources. Emotive confrontations about rights to the resource rents realised through corporate exploitation of the 'national patrimony' give place to less charged negotiations about appropriate levels of exported output, utilization of domestic components, and employment of nationals in managerial positions (although FDI in the service sector, particularly by US TNCs in the media and related industries, raises fears of diminished 'cultural sovereignty', not only in the Third World but in nations such as Canada). Overall, it has become increasingly clear that 'a bigger proportion of foreign investment is taking place in countries that want it and are well integrated with the world economy' (*The Economist* loc. cit.). The ideological stance of a government thus assumes considerable significance (see Fig. 5.2).

Since the 1930s, there has been a growing involvement by governments throughout the capitalist world in the functioning of their national economies. This has influenced not only the geography of economic activity within each state, in ways that are detailed in Chapter 8, but also the changing geography of global production. The nature of government intervention has taken different forms, with different international consequences. Perhaps the most far-reaching global impact has resulted from the distinctive character of government–industry relationships in Japan, which has encouraged the adoption of strategic long-term programmes of industrial growth and diversification geared to provide the large population of resource-poor islands with a high standard of living in a world of rapid technological change. This model of state-coordinated, but predominantly private-sector, industrialization is one which some of the Asian NICs (notably South Korea) are attempting to follow. There are specific cultural and historical features underlying Japan's rapid postwar attainment of a leading position among manufacturing–exporting nations, with all the consequences this has had for employment and corporate profitability in other core nations (Turner 1982b). But it would be misleading to attribute Japanese success purely to national characteristics: choices made by governments elsewhere have had considerable influence, albeit indirectly. For instance, national priorities in the USA (and to a lesser extent in the UK) have prompted governments to encourage and subsidize a concentration of R&D expenditures in electronics on military projects at the expense (despite some obvious 'spin-offs') of development in the consumer field (see Ch. 12). Turner (1982a, pp.57–8) argues, 'it is significant that the American industry left the Japanese and Philips [the Netherlands-based TNC] to develop video-recorders, for this must be the first major consumer product developed over the last twenty years in the development of which US companies have played no part.'

If the geography of production at a global scale increasingly reflects the differing capability of nations to adjust to a changing economic order, capabilities which are as much influenced by the evolution of distinctive political cultures as they are in the Japanese case, so too does it reflect the differing capabilities of corporations to respond to a dynamic external environment. To some extent, of course, the flexibility of nations and corporations is related, for TNCs are directed from their home states by personnel who do not (yet?) evidence an entirely homogenous global 'corporate culture' (Hamilton 1976, Negandhi & Baliga 1981). Moreover, the capacity of firms to adjust often depends on the degree of supportive or restrictive action by the states in which they operate.

Franko & Stephenson (1981, p.195) note that European, Japanese, and US TNCs 'seem to have developed distinctive strategies towards the export thrust of the NICs.' The typical European response has been to seek solutions to competitive pressures by 'rationalizing' output (see p.162) or shifting (through innovation) to higher-value-added production within a corporate geography that is largely confined to the core regions of Europe and North America. This strategy has still left too many firms vulnerable to 'losses of market share to the large-scale, low-cost, globally minded Japanese, whose productive operations typically [extend] into the NICs, and who [are] meanwhile themselves upgrading their own product lines and innovative capability' (ibid., p.196). Indeed, many Japanese firms are characterized by a willingness to *promote* change rather than simply to adapt to it, and they have actively 'cascaded' elements of the production process to lower-wage or less congested locations in Asian NICs (Dicken 1988). Franko & Stephenson (1981, p.196) suggest that the majority of 'US companies have often been caught in an uncomfortable middle position, being outdistanced both in process and product innovation, as well as doing little, compared to the Japanese, to seek export-production bases in the low-cost, fast-growth NICs.'

Both in Europe and North America, loss of market share at home has prompted some corporations to seek protection from their governments, incongruous though this may be with the fundamentally internationalist philosophy of multinational enterprise. Other firms have expanded their foreign operations while simultaneously contracting their business and employment 'at home'. Both courses of action illustrate the ambiguous relationship between the State and private enterprise in a mixed-capitalist economy, the geographical dimensions of which at the national and subnational levels are explored in the next chapter.

8 *The industrialized Western nations in a turbulent global economy*

Introduction

The greater global interdependence of economic activity since the Second World War has affected the industrialized capitalist nations to differing degrees and at different times. Economies such as the UK and the Netherlands have always been more 'open' (i.e. more dependent on trade) than those such as France and, especially, the USA. The relative self-containedness of the US economy made it easier for Americans to ignore the signs of their gradual loss of industrial hegemony, until major disruptions of the *status quo* in the 1970s (see Ch. 1) revealed just how far their economy was losing its competitiveness and had become integrated, vulnerably so, into the global economy.

Yet the impoverishing retreat into protectionism of the 1930s had persuaded all the developed capitalist nations to work for a freer international trading regime after 1945, and the cumulative rounds of tariff reductions in the 1950s and 1960s gave this substance. The benefits of faster economic growth were widely shared, even in laggard performers such as the UK. What turned the tide of sentiment towards the neo-mercantilistic policies of the 1970s and 1980s was the emergence of strong, 'disruptive' competition from Japan and the Asian NICs, which magnified the domestic problems the older industrial economies of Western Europe and North America were having in coping with accelerating structural change. The changing economic geography of these nations during this turbulent era is analysed in this chapter.

Industrial location and regional economic development

The origins of 'regional policy'
After 1945, governments of the industrialized Western nations adopted Keynesian policies to regulate their national economies and maintain full employment. Significant and persistent *regional* variations in unemployment seemed to require geographically selective policies to contain disparities within politically acceptable limits (Armstrong & Taylor 1978). Such 'regional policies' had actually originated in the Depression of the 1930s. Federal initiatives in the USA had included the creation of the Tennessee Valley Authority to promote a broad set of social and economic development objectives in southern Appalachia (Friedmann & Weaver 1979). The British government designated four coalfield industrial regions suffering from severe unemployment (over 50% in some localities) as 'Special Areas', eligible for (limited) financial assistance to support business expansion (McCrone 1969). The Prairie Farm Rehabilitation Act was a Canadian government response to rural poverty, exacerbated by depressed grain prices and environmental degradation. These various measures shared the characteristic of being essentially *ad hoc* responses to regionally concentrated social and economic distress. Following 1945, except in the USA, regional policies of a much more comprehensive nature became popular.

The UK experience
The UK has pursued regional policy measures over a longer continuous period and with a greater level of effort than any other state. The symptoms of an emergent core–periphery contrast in economic prosperity were identified with great precision by a Royal Commission in 1940. The coalfield industrial areas such as north-east England, South Wales, and the Clyde Valley were visibly in decline, whereas London and other cities in south-east England had attracted new growth industries even during the Depression. Concern about growing congestion and loss of agricultural land to industrial and urban expansion in south-east England was reflected in proposals to encourage the dispersal of industry to other regions. This would also ensure, it was argued, that the social and economic infrastructure of less prosperous regions was not under-utilized. Hence a combination of physical planning legislation, which restricted industrial development in the national core, and an expanded programme of assistance to new industrial establishments in the peripheral 'Development Areas' was used to attempt to bring about a better regional balance of growth and employment (McCrone 1969, Hall *et al.* 1973; Fig. 8.1).

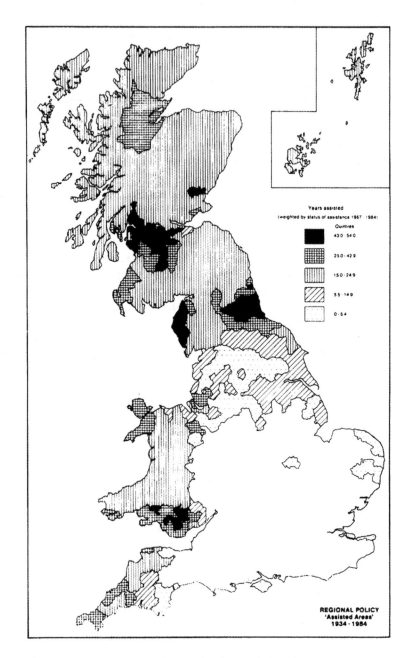

Figure 8.1 British regional policy: assisted areas 1934–84.
Source: Townsend, A.R. 1987. Regional policy. In *Industrial change in the United Kingdom*,
W.F. Lever (ed). Harlow: Longman; Figure 14.1.

One can see in retrospect that British policy-makers never probed sufficiently the question of how far 'regional' problems were truly *regional* (Massey 1979). The severe unemployment which had become a permanent feature of coalfield industrial areas reflected changes both in the *structural* (sectoral) composition of the national economy and in the UK's rôle within the capitalist world-economy (Dunford & Perrons 1983). The growth industries of the 1920s and 1930s embodied technologies based on oil or electricity rather than coal, and were predominantly 'light', consumer goods industries (including automobiles) rather than 'heavy', capital goods industries. Moreover, the disruption to world trade caused by the First World War had encouraged the development of manufacturing in Third World nations such as India and Argentina, so that British exports, especially cotton goods, permanently lost a substantial share of their overseas markets. Because of the pronounced geographical concentration of output in the industries on which Britain's 19th-century prosperity had been based, structural change in the national economy translated directly into rising unemployment in specific regions. The problems faced by areas such as South Wales were exacerbated by the fact that the composition of their industry and employment was so undiversified.

British regional policy was targetted on areas defined by persistent high unemployment. Given unprecedented national economic growth and continuous job-creation in the manufacturing sector in the 1950s and 1960s, it was assumed that the preference of industrial decision-makers for locations in the Midlands and South-East could be countered, with little risk to corporate profitability, by government measures to induce expansion in the peripheral regions. British policy thereby expressed the conviction that it was politically, socially, and probably economically (although this was very hard to measure) preferable to 'take work to the workers' rather than to encourage migration from peripheral to already congested core regions. Indeed, one of the economic arguments for State intervention was that macro-management of the national economy, especially the control of inflation, would be easier if the demand for scarce factors of production (notably skilled workers and land) in the prosperous regions could be moderated (McCrone 1969).

After 20 years of increasingly comprehensive and generous regional incentives, the map of manufacturing employment in the UK had been significantly changed by the early 1970s (Keeble 1976). The Development Areas contained a more diversified, modernized, and larger industrial base than they would have done in the absence of government assistance. Improved infrastructure, especially the construction of a national motorway network, also contributed to their fuller integration into the more prosperous economy of the national core region. Nevertheless, it became increasingly apparent that the convergence between core and periphery, as measured by unemployment rates for instance, was no longer just the result of positive steps to improve the position of disadvantaged regions; it also reflected stagnation or rising unemployment in areas which until recently had been the centres of growth. This raised more acutely the question of how far 'regional' problems are location-specific, and of the effectiveness of policies which treat the geographical distribution of industrial employment in isolation from the institutional (corporate) organization of production processes (Massey 1984).

Shifting regional prosperity in the United States
Disparities between average per capita income in different regions of the USA have shrunk steadily for a century. In 1880, the richest part of the country, the West, had almost twice the national average income, whereas the South had only half, and the North-East had 1.4 times the US mean. By the late 1970s, however, the richest and poorest regions were separated by only 25 percentage points, with residents of the South East receiving 86% of the national average income and those of the Far West 111% (Price 1982). This dramatic convergence of regional incomes came about with almost no *overt* action by the US federal government to promote regional equality. It has therefore been used by critics of regional policy in other nations to suggest that, left to themselves, market forces will even out regional disparities quite effectively. The limited programme of regional aid which *was* introduced in the USA in the 1960s met with the 'indifference, or even hostility, of successive federal administrations' (Estall 1982, p.35).

Yet there is more to a government's influence on regional economic wellbeing than so-called 'regional policy'. The geographical incidence of per capita spending by functional agencies or departments is unlikely to be even across any nation; and major shifts in the relative prosperity of different regions in the USA which became apparent in the mid-1970s (see p.163) prompted many commentators to identify the significant rôle of the spatial incidence of federal expenditures:

> Toward the Northeast [the federal government] offers a rigid, restrictive policy, taxing wealth out of the region's economy, depleting consumer buying-power. Toward the Sun Belt, it offers a policy of massive pump-priming and deficit spending, taking far less in taxes than it puts back in spending (Morris 1978, p.111; quoted by Browning 1981).

Available data indicate that the impact of government outlays was not so sharply regionalized as this partisan assessment implies, although the states in the lowest quintile were clearly concentrated in the traditional

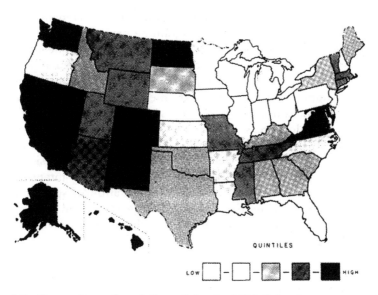

Figure 8.2 Per capita outlays (adjusted) by the USA federal government, by state, 1976.
Source: Browning, C.E. 1981. Federal outlays and regional development. In *Federalism and regional development: case studies on the experience in the United States and the Federal Republic of Germany*, G.W. Hoffman, (ed.). Austin: University of Texas Press; Figure 5-1.

Manufacturing Belt (Fig. 8.2). Not all forms of expenditure have the same multiplier effect, however, and there is little dispute that many of the more stimulative federal outlays in the 1960s, including the space programme and many infrastructure projects, were relatively concentrated in the West and South (Browning 1981).

Canada's persistent problem
Significant and persistent regional disparities in Canada date from at least the 1920s, when the incipient industrialization of the Maritime provinces was stalled by increasingly strong competition from Ontario and Quebec (Wynn 1987). A wide variety of unrelated programmes was introduced by the federal government during the 1960s in an effort to narrow the variation in unemployment and average income across the country (Brewis 1969). These were given a more coherent focus by the creation of the Department of Regional Economic Expansion (DREE) in 1969. Although accounting for barely one-third of DREE expenditures, the programme of financial assistance to private industry was the most geographically specific, and also the most controversial, component of Canada's attempts to improve the long-term economic health of its lagging regions (Economic Council of Canada 1977).

Modelled very much after British measures to attract manufacturing to peripheral areas, the Canadian programme nevertheless seemed not to learn from British mistakes. So extensive were the areas deemed eligible for assistance that by 1971 they contained half the national population, with resultant dilution of the programme's impact on regions most needing support. At the other extreme, many of the 'Special Areas' targetted for additional infrastructural support were far too small to possess growth-pole potential. The DREE industrial incentive programme was variously criticized for its limited *incrementality* (support for genuinely 'new' jobs), its irrelevance to corporate location decision-makers, and its tendency to favour capital- rather than labour-intensive investment (Walker 1980). Nevertheless, the application of specific regional policy measures during the 1960s and 1970s was credited with some part in narrowing the prosperity gap between the poorest provinces and the national average.

The demise of regional policy
The purpose and effectiveness of regional policies which had largely consisted of measures to promote industrial expansion in economically depressed areas were understandably called into question by the stagnation, then rapid decline, of manufacturing employment in many core nations in the 1970s (Chisholm 1976, Bartels & van Duijn 1982). Disillusionment has reflected a number of distinct concerns. There is, first, the difficulty of measuring unambiguously the results of established programmes. By the late 1970s, British regional policy had been in continuous effect for 30 years, and a number of analyses, based on different methodological procedures, attempted to quantify its achievements. Their results, reviewed by Frost & Spence (1981), exhibited considerable variation in the imputed magnitude of job-creation.

A second cause for hesitation was the feeling that, when manufacturing employment growth nationally is negligible or negative, regional policies do no more (at considerable expense) than shuffle the geography of unemployment. The emergence of a significant volume of unemployment in regions which for decades had been discriminated against, such as the West Midlands, certainly hastened the reconsideration of British policy (Flynn & Taylor 1986).

A third development which challenged the relevance of place-specific measures to restore or maintain manufacturing jobs was the growing volume of financial assistance which governments felt obliged to channel into specific firms or industrial sectors. It is, after all, corporations which become internationally uncompetitive and face bankruptcy, rather than regions as such. Although there is an implicit regional dimension to all aid granted to firms (reflecting the particular corporate geographies of such recipients as Lockheed, Chrysler, and the former British Leyland and Massey-Ferguson), State support usually has as much to do with concern for national manufacturing capability or trading prospects as with the domestic distribution of employment. The more that sub-ventions to industry take on a sectoral or corporate, rather than a regional, dimension, the more questions arise about whether the money would be better spent backing 'winners' than 'losers' (few politicians are prepared to identify regions quite so categorically!). By the early 1980s, governments in the UK, Canada, and other core nations showed more interest in removing legislative and bureaucratic obstacles to industrial expansion, wherever it appeared to be taking place, than in persisting with policies to channel it to traditionally depressed regions.

A fourth argument for reconsidering regional policy was that it had remained fundamentally geared to stimulating manufacturing employ-ment despite the fact that, in all the advanced industrial economies, it is the service sector that employs the majority of the labour force and has shown the most dynamic growth since 1945. The persistent emphasis on manufacturing reflected the belief, which is not entirely misplaced, that this sector offers the greatest proportion of jobs which are potentially mobile at an interregional scale (although the 'quality' of jobs in branch plants established in peripheral regions has come in for critical scrutiny). Specific attention to the distribution of service employment had been essentially confined to Britain and France, where, for a period in the late 1960s, it took the form of restraints on office expansion in London and Paris. These negative measures arose more out of a desire to deconcentrate activity away from the capital cities than from a well thought-out analysis of the rôle of service sector activity in maintaining or redressing regional economic disparities. Indeed, McEnery (1981) argues strongly that government attitudes which consistently denigrated the economic importance of 'producer services' have resulted in Britain in their over-concentration in the South East and the reinforcement of core–periphery contrasts, the sociological dimensions of which are more damaging to the revival of prosperity in depressed regions than their immediately obvious economic ones.

The fifth challenge to regional policy was that, when growth in the national economy falters, governments cannot afford to intervene in the geographical distribution of employment on a scale that is likely to have much impact. Together with the disappointing experience of many regions, where programmes to attract non-local manufac-turing firms resulted in disproportionately few permanent jobs for the effort and expenditures involved, fiscal constraints have promoted a wider recognition of the importance of local initiative and a realistic appraisal of indigenous resources in the process of regional revitalization (Segal 1979). The learning process reflected in the changing emphases of the Cape Breton Development Corporation in Canada since its creation (George 1981) has involved acknowledging the limits both of 'home-grown' development and of industry which is 'parachuted in' to a regional economy that is ill equipped to sustain it. In a period when neither spontaneous increases in employment nor substantial State assistance can be expected, cities and regions in core and periphery alike resort to increasingly sophisticated publicity to sell themselves to potential new employers (Burgess 1982).

The final nail in the coffin of postwar regional policy has been the revival of a dominant political philosophy which views State intervention in the marketplace, particularly to back perceived 'losers', as anathema (Chisholm 1987). This has been most thoroughly implemented in the UK since 1979, under the Thatcher administration, significantly reducing the geographical coverage and financial benefits of regional support measures (Martin 1986).

The changing industrial economy

By the late 1950s, the postwar industrial hegemony of the USA was being challenged by the renewed manufacturing potential of Western Europe. A decade later, US and European firms were both losing ground in crucial sectors to growing Japanese competition. By the mid-1980s, selected industries in Japan were themselves feeling the competitive edge of NIC producers. In other words, the rapid postwar growth of core nations was taking place in an increasingly exacting business environment even before the international economic dislocations of the 1970s put a premium on the *flexibility* of national and regional economies to respond to the changing currents of global comparative advantage. In practice, the ability of the State, at various geographical scales, to adjust to a new economic environment has been increasingly circumscribed by the growth of TNCs capable of organizing production on a worldwide, supra-national basis, and by the (varying) political successes of the labour movement and regional interest groups in protecting themselves against the immediate consequences of declining international competitiveness.

The steady growth of employment in the service sector of core industrial economies, which had caused the *relative* decline in the size of the manufacturing sector, tended, until the early 1980s, to mask significant changes which have taken place in the relationship between industrial production and manufacturing employment (Rothwell 1982). As indicated in Figure 8.3, until the mid-1960s, rapid growth in output in Western Europe was associated with a slower but steady rise in employment. As a result, the *absolute* size of the manufacturing

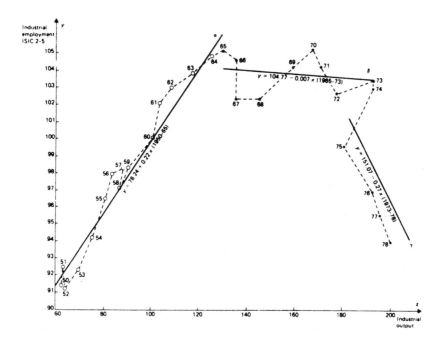

Figure 8.3 Industrial output and employment in the nine-member EEC, 1950-78 (1960 = 100).
Source: Rothwell, R. 1982. The role of technology in industrial change: implications for regional policy, *Regional Studies* 16; Figure 1.

workforce continued to grow. After 1965, however, manufactured output maintained its expansion without creating additional employment, a change which was also noted in the US Manufacturing Belt (Norton & Rees 1979). In retrospect, this development can be seen to have coincided with a rise in the 'underlying' level of unemployment, in the UK in particular (Massey & Meegan 1982). The relationship between output and employment in the manufacturing sector changed again after 1973, when a much reduced (but still positive) rate of growth in production was accompanied by substantial absolute losses in employment. The volume of job losses increased greatly in the deep recession of the late 1970s and early 1980s, as output contracted and talk of the 'de-industrialization' of the core economies of the West suddenly became credible (Blackaby 1978, Bluestone & Harrison 1982, Townsend 1983).

Massey & Meegan (1982) analyse 'the anatomy of job loss' within a national economy of declining overall competitiveness (the UK) in terms of three processes of corporate adjustment. *Intensification* is an approach pursued by firms aiming to increase labour productivity with a minimum of new investment. *Investment and technical change* is a strategy which relies on substantial expenditure to introduce labour-saving production technologies. *Rationalization* is a straightforward reduction in productive capacity and the jobs which go with it. The lack of a direct link between changes in output and employment is apparent, for the first two adjustment processes can conceivably cause them to move in opposite directions, with production actually rising but the workforce shrinking. The geographical consequences of these different strategies of adjustment are by no means immediately predictable, either. Whereas intensification takes place, by definition, at existing job locations, and rationalization involves the elimination of *some* of a set of existing establishments (with perhaps some redeployment of work to others), investment in technical change necessarily brings with it the possibility that new locations for production will be chosen. Taken together, the processes of corporate adjustment to conditions of slow economic growth or decline, and the continuing (albeit more limited) emergence of expanding firms associated with new technologies means that the map of industrial production is far from static. Indeed, some quite dramatic shifts in the regional distribution of manufacturing in the industrialized core economies have taken place since the early 1970s.

At the regional scale, the principal development has been the decisive change of fortune of the previously prosperous concentrations of manufacturing activity. Declining international competitiveness, as felt by the automobile and steel industries in the USA and the UK, for instance, disproportionately hit employment not simply in 19th-century centres of industry, such as Pennsylvania and North-east England, but mid-20th century centres such as the Detroit region and the West Midlands (Townsend 1983). Even California had, by the mid-1970s, a mix of industries which reduced its growth prospects below the national average (Rees 1979). At the urban and metropolitan scale, the most obvious feature is the poor showing of large industrial cities and, conversely, the substantial gains in industrial employment in small towns and rural areas. The loss of manufacturing jobs in inner-city areas has been particularly noticeable, for corporate rationalization has consistently made the deepest inroads into the relatively skilled labour force traditionally found there (Massey & Meegan 1978, Scott 1982). In contrast, manufacturing employment in small urban centres in regions with little or no manufacturing tradition has grown substantially, throughout much of the southern USA (Lonsdale & Seyler 1979) and in parts of rural England, notably East Anglia and the South West (Keeble 1980).

These geographical shifts reflect a wide variety of underlying influences, many of which have been operative for decades, others of which are clearly of quite recent origin. Industrial production in capitalist countries has still to meet the ultimate test of profitability, and newly emerging locational patterns are a response to this imperative in a world which has changed significantly since the 1950s. In the USA, for instance, the erosion of the traditional industrial core region's dominance in manufacturing output and employment began in the 1920s, with the loss of jobs in the textile and clothing sectors to cheaper labour markets in the South. Corporate adjustment to regional wage differentials became much more pronounced in the 1960s, as the increased mechanization of routine production processes in a growing range of industries made it possible to draw upon relatively unskilled labour for jobs associated with the mature phase of the product life-cycle (Norton & Rees 1979). Competitive pressures within an increasingly global, not merely continental, market have resulted, however, in a significant loss of low-skill jobs from regions of recent manufacturing growth to Third World locations (Peet 1983).

Compounding this process of the diffusion of industrial employment from the core to the periphery of the USA (so that by the mid-1970s the Manufacturing Belt's share of jobs in manufacturing fell below one half for the first time ever) was a shift in the relative significance of the two regions' resource endowments and transportation potential. The bases of the Manufacturing Belt's early-20th-century prosperity (coal, iron, and steel, a dense railway network, and a clear majority of the national population) have been increasingly eclipsed since 1945 by those of the southern and western periphery (oil and gas, petrochemicals, a good highway network, and a population increasing in size and affluence). These changes have been augmented by cultural and environmental factors, notably the improved race relations and near-universal air-conditioning of the South. Specific developments in the 1970s further reinforced changes in the balance of economic advantage between core and periphery. More open than ever before to the currents of international trade, the USA has responded to its declining competitiveness as a source of manufactured goods and the heightened insecurity of global energy supplies by stimulating agricultural exports from the Great Plains and tapping its vast coal reserves in the West. The cumulative effect brought about what some claimed was an unprecedented 'principal axis shift in the American spatial economy', with the Gulf coast replacing the Northeastern seaboard as the focus of activity (Vining et al. 1982).

However, it is symptomatic of the increased instability of the contemporary world-economy that by the mid-1980s, a radically different map of regional prosperity was identified in a Congressional committee report. 'The Bicoastal Economy' (actually, California plus the Atlantic coast) experienced per capita growth (1981–5) three times greater than the central (and Pacific North West) USA, where the price decline of agricultural commodities, energy, and other resources brought severe economic hardship (The Economist 19 July 1986). In fact, throughout the swings in the popularized geographies of regional change ('Sunbelt/Snowbelt', etc.) since 1970, and despite the diffusion of manufacturing into the South and West, the control of US industry has remained overwhelmingly concentrated in the corporate headquarters of the old Manufacturing Belt (Birch 1979). The dynamism of southern California reflects a particular mix of favourable conditions, reviewed below.

The geography of growth in high-technology industries

Much of the rapid economic growth of the 1948–73 era was based on the perfection and widespread diffusion of already known technologies, such as the internal combustion engine and industrial applications of organic chemistry, rather than on fundamentally new technologies. The growing competitiveness of a wide range of 'mature' manufactured goods from nations without an industrial tradition, such as the NICs, has re-emphasized the degree to which the continued prosperity of the core industrial states depends on their maintaining a superior capacity to introduce technological innovations. With successful manufacturing performance becoming increasingly dependent on organized R&D, and with the military imperatives to harness new technologies assuming an unprecedented momentum, the economic and strategic significance of applied science in the core nations is indisputable. In some industrial sectors there is a complex interaction between civilian and military research, well illustrated by the innovations associated with the US space programme in the 1960s. But nations such as the US and the UK, whose R&D spending involves a substantial defence-related component, have been shown to derive less economic benefit from these outlays than countries such as Japan and West Germany, whose R&D focus is more directly attuned to potential markets for sophisticated capital equipment or consumer products (see p.150).

It is significant that the technologies which have become, or are becoming, critical for industrial leadership in the contemporary world make far fewer demands on the Earth's material resources than did the 'coal, iron and steam' technologies of the early Industrial Revolution or even the 'oil, chemicals and electricity' technologies of the 1950s. The growth of scientific knowledge about the basic building blocks of the physical universe and its life forms has made it increasingly possible to harness the energy and order of natural phenomena at the molecular level. Microelectronics and genetic engineering are prominent among the bases for radically new technologies which have begun to appear

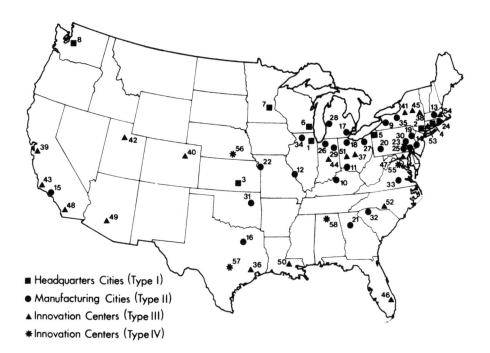

■ Headquarters Cities (Type I)
● Manufacturing Cities (Type II)
▲ Innovation Centers (Type III)
✳ Innovation Centers (Type IV)

Figure 8.4   R & D location types among 58 US urban areas.
Source: Malecki, E. J. 1980. Corporate organization of R & D and the location of technological activities, *Regional Studies* 14; Figure 3.

since the early 1960s. Industries built around these technologies have very different characteristics, reflected in their locational preferences, from those which have dominated the core economies until fairly recently (Hall & Markusen 1985).

Research and development laboratories and related 'high-technology' manufacturing establishments exhibit distinct geographical concentrations. In core–periphery terms, they are clearly concentrated in core regions: within these, they are noticeably concentrated in and around the largest metropolitan areas. The reasons for this distribution are twofold. The rationale for locating R&D activity and the early production of innovative goods in close proximity to corporate headquarters was set out in Chapter 6, and similar considerations in the public sector favour locations close to central-government agencies. This tendency for employers to promote the agglomeration of R&D establishments in a restricted number of major metropolitan areas has its counterpart in the decision-making of the highly trained and geographically mobile personnel who are the principal resource of these institutions. Preferring an environment of high cultural amenity, professional stimulus, and

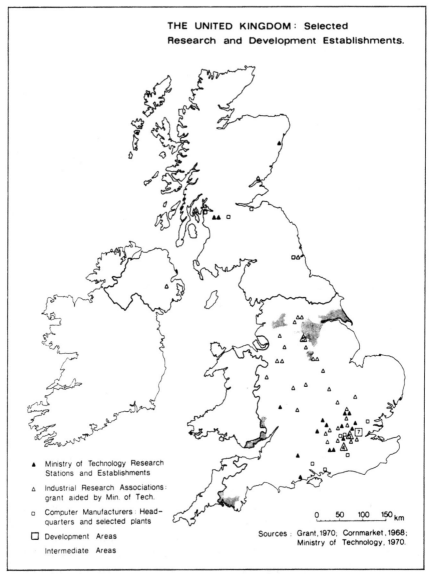

THE UNITED KINGDOM: Selected
Research and Development Establishments.

▲ Ministry of Technology Research
   Stations and Establishments

△ Industrial Research Associations:
   grant aided by Min. of Tech.

□ Computer Manufacturers: Headquarters and selected plants

☐ Development Areas

   Intermediate Areas

0   50   100   150 km

Sources : Grant,1970; Cornmarket,1968;
          Ministry of Technology, 1970.

Figure 8.5   Selected R & D establishments in the UK, 1970.
Source: Hall, J. M. 1970. Industry grows where the grass is greener. *Area* 2(3); Figure 1.

good prospects for career advancement, employees are most attracted
to the established centres which can offer them.

   Malecki (1980) identifies four categories of R&D centres among 58
US cities which contain at least five laboratories (Fig. 8.4). Eight cities,
ranging in size from New York to Wichita, Kansas, contain clusters of
research establishments clearly associated with the location of corporate

headquarters. Almost half the total of cities form a second category, whose attraction of R&D facilities is mainly a function of their importance as manufacturing centres. The next largest category is a group of cities which Malecki (1980, p.231) terms 'innovative centres', on the basis of their being 'university cities with some federal scientific activity and some manufacturing'. His final group is a set of four cities which are extreme examples of the previous category; centres with few corporate headquarters and little manufacturing activity, but with very substantial involvement in university-based research or in-house R&D by federal agencies.

Buswell & Lewis (1970) and Howells (1984) have analysed the distribution of industrial research activity in the UK (Fig. 8.5), and it is instructive to compare their findings with those of Malecki (1979, 1980, 1982). The USA and the UK are similar in displaying core–periphery contrasts in the concentration of R&D facilities. Significantly, in both countries, university-based research contributes particularly to geographical dispersal. On the other hand, the USA differs from the UK in possessing an unusually large number of R&D locations. This is more than simply a reflection of the greater size of the US economy. It results from differences in the urban systems of the two nations and in the proportion of R&D carried out by private corporations as opposed to that performed in government establishments. London's clear dominance as the political and economic capital of the UK and the large proportion of research in Britain which is publicly funded has promoted an extreme concentration of R&D facilities in south-east England which is unmatched in the USA. There, the dominance of New York as the R&D capital has declined noticeably since the mid-1960s, at the expense of other large metropolitan areas, particularly Los Angeles. At the same time, the vulnerability of research centres that are heavily dependent on federal funding to changes in national political priorities has been evident in the decline of NASA-related establishments in the South. Implications of these trends for regional prosperity are discussed further below.

How far has growing employment and output in R&D-intensive industries offset declines in traditional manufacturing sectors? It is not easy to give a definite answer: in addition to the problems of distinguishing cyclical from structural factors in interpreting the performance of national economies since the mid-1970s, one faces the conceptual challenge of correctly identifying the configuration of the production process as commodities become increasingly knowledge- rather than material-intensive. There is no uniform definition of 'high-technology' industry, and any meaningful one must bridge the conventional distinction between the manufacturing and service sectors. Manufacturers of computer hardware and compilers of computer software (classified as 'producer services' if

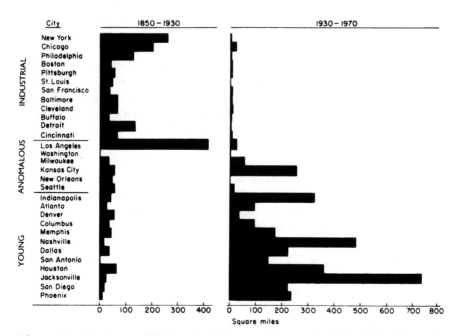

Figure 8.6 Territory added by the 30 largest cities in the USA, 1850–1930 and 1930–70. Note that annexations by old industrial cities were generally modest and all but ceased after the 1920s, whereas young metropolitan centres have been able to accomplish substantial territorial expansion in recent decades.
Source: Norton, R. D. 1979. *City life-cycles and American urban policy.* New York: Academic Press; Figure 4.2

they are independent business firms) both belong (see Fig. 6.5). Some insight into the dynamics of employment change can, however, be gleaned from an intensive study of the New England economy, part of which is reported by Harrison (1982).

New England differs significantly from most other regions where traditional industries are in decline by virtue of possessing, in the metropolitan Boston area, a leading centre of innovation. Even with this advantage, however, the transition away from a specialized employment base strongly fashioned by 19th-century mill industries (textiles and clothing, shoes, and paper) and mid-20th-century growth in the mechanical and electrical engineering sectors has not been an unqualified success. Although a number of high-technology industries have expanded very rapidly, especially those associated with computer manufacturing and scientific instrumentation, the loss of jobs in traditional manufacturing activities has been made up only by a massive expansion of service-sector employment, concentrated in retailing and distribution, health care, and the 'hospitality' industry. Harrison discerns a number of features in the evolving employment pattern which must temper one's optimism about the region's prospects:

> More people are working than ever before. But the wages paid in a growing share of these jobs are declining . . . in real terms . . . . Jobs in unstable (e.g., part-year) industries are expanding, while employers are reorganizing work in ways that exacerbate that instability . . The best-paying jobs, in the engineering industries (and, more recently, in the production of computers as well) are becoming subject to sharp 'boom–bust' cycles linked to rapidly changing technology and fluctuating foreign and domestic government military procurement. The distribution of earnings is becoming much more unequal over time, within and among industries, . . . between the races, and by gender (Harrison 1982, pp.41–2).

This evidence, from a core region of a global core economy, suggests that employment in both the manufacturing and service sectors is taking on an increasingly polarized character. One can identify 'good' jobs and 'bad' jobs, in terms of pay levels, hours of work, and the stability and security of employment. The intermediate layer of semi-skilled manufacturing jobs and of middle-management positions, in retailing for instance, is contracting sharply. The New England experience warns us that even in core regions which are the corporate home of high-technology industries, high-technology machine tools permit a wide range of fabrication processes to be undertaken with limited quantities of relatively unskilled labour. Similarly, in many parts of the service sector (especially retailing and the 'hospitality' industries), contemporary information-gathering and -processing technology permits decision-making to be concentrated amongst a restricted number of employees at key locations, so that the numerous positions involving delivery of the 'product' to the customer become relatively unskilled jobs which can increasingly be confined to minimum-wage employees. Soja, Morales & Wolff (1983) note these same processes of labour-force polarization in another centre of 'high-technology' manufacturing, the Los Angeles metropolitan area.

There are core–periphery contrasts in the geography of high-technology manufacturing in the UK, although 'Silicon Glen' in central Scotland has a notably higher level of research activity and knowledge-intensive employment than other peripheral regions. The M4 Corridor, west of London (Breheny et al. 1985), and the Cambridge region (*Financial Times*, 12 February 1985) have emerged as agglomerations within a decentralized pattern of high-technology industry in the South East. Both areas are characterized by the active formation of small firms by innovative entrepreneurs, many of whom have been 'incubated' within the University (Cambridge) or government research establishments (M4). (Steed & de Genova (1983) document a Canadian example of this process in the Ottawa region.) In contrast, employment growth in microelectronics plants in South Wales owes almost nothing to firms indigenous to the region. Unskilled women have filled the majority of positions created by new branch plants of (mainly foreign) firms, producing standard rather than customized equipment (Morgan & Sayer 1985).

Population shifts and the changing fortunes of metropolitan areas

The changing geography of 'place prosperity'
Since the early 1970s, changes have become apparent in varying degrees in all the Western core nations which have qualified, but not eclipsed, the analytical value of broad *regional* distinctions between healthy and ailing economic environments. In particular, core–periphery contrasts within *daily urban systems* (metropolitan labour markets or 'commutersheds') have sharpened, especially where the centres of these systems are large metropolitan areas. The inner city is now seen, in Western Europe as much as in the USA, to exhibit an economic malaise which is as deep-seated as that of any region whose traditional industries are in structural decline (Thrift 1979). Many rural (non-metropolitan) areas have, in contrast, improved their prospects considerably. Among smaller industrial cities, levels of economic activity have become more varied, largely reflecting the fortunes of particular establishments within corporate strategies of restructuring. The net effect is that 'the spatial incidence of relative growth and decline has become distinctly more localised, the greatest disparities now being at the level of the "district" rather than the broader regions' (Martin & Hodge 1983, p.142).

Shifts in the geographical distribution of population are a sensitive barometer of the changing pattern of 'place prosperity'. The geography of employment remains the single most important influence on where people live, and the changing spatial preferences of employers, both inter- and intraregionally, are responsible for much of what is novel in the contemporary population patterns of core nations. Nevertheless, non-employment income is of growing significance: its share of total personal income increased from 18 to 29% in the USA between 1950 and 1975 (Beyers 1979). One geographical expression of this development is that it permits many people of retirement age to choose a place of residence on the basis of preferences for specific social and/or natural environments. Maps of national population distribution have thus come increasingly to reflect a complex mix of perceived economic opportunity and residential attractiveness. Those who are free to move provide mounting evidence that traditional industrial regions and inner metropolitan areas are seen to offer neither.

Internal migration flows are the key factor behind evolving regional contrasts in population growth rates. A substantial decline in rates of natural increase took place in all the Western core states in the 1970s, to the extent that some European regions with a negative migration balance can no longer rely on natural increase to cover their losses (Bartels & van Duijn 1982). In addition, growing domestic unemployment has prompted nations to reduce the flow of international immigration. By the early 1970s it became clear that many large cities face sizeable and persistent reductions in their resident population. The emergence of 'circular and cumulative' processes which reinforce decline rather than growth has created serious economic difficulties for metropolitan administrations. The loss of businesses and the disproportionate departure of high- and middle-income households create a situation in which not only do taxes rise while standards of *public* services deteriorate, but the traditional sources of *private* philanthropy decline just as their contribution to the welfare of the city becomes ever more necessary. (Wolch 1983).

Job and income loss in inner cities
One element of the continuity of recent changes with long-established trends is the erosion of blue-collar employment opportunities in the central (or inner) cities of the largest industrialized metropolitan areas (Lever 1981, Scott 1982). Manufacturing jobs have been in decline since the 1950s, but a dramatic acceleration in the rate of loss became evident in the 1970s. New York City, for instance, having lost an average of 11 000 factory jobs a year between 1950 and 1969, lost an average of 43 000 annually in the period 1969–75 (Tabb 1978). Inner London lost just over one-fifth of its manufacturing employment (135 000 jobs) between 1966 and 1974 (Dennis 1978). The diminishing attractiveness of inner cities as manufacturing locations has been evident since the 1920s, but the rapid deterioration of their position in recent years reflects the increasingly internationalized global division of labour. Although the residents of poor-quality housing in inner-city neighbourhoods have traditionally been an abundant source of cheap labour in core economies, many of the firms employing them, in sectors such as the clothing industry, have fallen victim to the competition of even lower-wage producers in Third World countries (Steed 1976, 1981). In addition, the skilled and semi-skilled blue-collar jobs associated with central-city engineering plants dating from before the Second World War have been particularly prone to shrinkage in the process of corporate rationalization of excess manufacturing capacity (Massey & Meegan 1978).

Service-sector employment in many inner cities has not grown sufficiently to offset losses in manufacturing. Office employment in the central business district of the largest US metropolitan areas was generally increasing until the mid-1960s. Then automation and/or the transfer to suburban or smaller city locations of routine administrative units began to stem this growth. The departure of corporate headquarters from sites in New York City (Quante 1976) added a new dimension to the declining attractiveness of old central cities in the USA and, by the 1980s, this trend had become evident even in London, which overwhelmingly dominates the British corporate hierarchy. Meanwhile, blue-collar service employment in inner-city areas was declining steadily. Study of the 'depot economy' has been rather neglected, but it is clear that technological change in, say, the transport industries (from steam to diesel traction on the railways, and from traditional cargo-handling to containerization in the ports), the energy sector (the decline of coal and gas), and varied maintenance operations has drastically cut the volume of employment (Townsend 1977). Wholesaling (and in the USA at least, retailing jobs also) has declined. Overall, job losses in inner cities have been at the expense of those lower socio-economic groups who have traditionally been constrained to reside there, whereas employment expansion has generally benefitted the higher socio-economic groups who commute from outer suburbs or who have colonized gentrified central neighbourhoods (Smith 1982).

Particularly within the geopolitically fragmented metropolises of the US Manufacturing Belt, declining central-city employment has been the principal cause of acute fiscal crises:

> Job loss is inevitably followed by tax-revenue loss . . .. If the half million jobs that disappeared [from New York City] between 1969 and 1975 were today providing income for New Yorkers, the city would be receiving $1.5 billion more in tax revenues and there would be no crisis (Tabb 1978, p.246).

City budgets are squeezed from both ends, for while the tax base contracts, demands for expenditure do not. Sternlieb & Hughes (1975, p.2) catalogue some of the factors with which city authorities have to contend, including the aging infrastructure of utilities and transportation systems 'which now represents very substantial capital and operating costs but, given the decline in usage, becomes much more a hindrance than a positive asset.' Friedland (1981) notes that among the consequences of deferred investment in New York City are the leakage of half of the throughput of the water supply system, and a replacement cycle for road surfaces of 150 years (as against a 25-year optimum).

The plight of central cities in the industrial metropolises of the northeastern USA demonstrates the intimate connection between the dynamics of change at the national and the metropolitan scales. The North East has been experiencing a net loss of whites through outmigration since 1940, but until 1970 this was more than counterbalanced by the net in-migration of blacks, mainly rural workers from the South. Black newcomers became overwhelmingly concentrated in the central cities, which non-migrant whites were steadily deserting for the suburbs (Sternlieb & Hughes 1975). The experience of inner-city blacks in the 1960s and 1970s failed to match that of earlier occupants of the metropolitan core, for reasons that are both economic and sociological. Whereas generations of newcomers ('ethnic', but white) prior to the Depression had found the low-wage economy a launching pad to upward socio-economic mobility, blacks were prevented by poverty and various manifestations of racial prejudice from leaving the ghetto, at a time when low-wage jobs in the manufacturing and service sectors were on the decline. The rising level of un- and underemployment faced by this group (and more recently arrived minorities) has imposed growing welfare burdens on city budgets that are increasingly unable to sustain them. In the North East, 'urban poverty has become a symptom of decline ... because it is no longer viewed as either economically or politically manageable' (Perry & Watkins 1977, p.279).

In contrast, poverty, or relative poverty, in the South is usually interpreted in a more positive light. Low wages represent a 'good business climate' and the scale of public spending on welfare measures is not yet regarded as parasitic on wealth creation in the private sector. These contrasted perceptions are part of the popular but oversimplified (and decreasingly useful) distinction between the declining 'Snowbelt' and the prosperous 'Sunbelt' (Browning & Gesler 1979, Mollenkopf 1981).

There are indeed critical differences between the political economy of the old industrial metropolises of the Manufacturing Belt and of the newer cities of the South, which temper social and economic polarization between central and outer areas in the latter (Fig. 8.6). Nevertheless, the South (especially the South East) is not immune from the pressures felt in the North. Between 1972 and 1977, the rate of decline in central-city manufacturing employment in Miami and Atlanta was only marginally less than in New York (Scott 1982). On the basis of a comprehensive measure of 'subemployment', Perry & Watkins (1977) found that inner-city populations in the Sunbelt are, if anything, slightly more disadvantaged than those in the North East. Part of the continuing economic dynamism of Los Angeles (whose experience of ghetto riots in the late-1960s matched that of any northeastern city) is attributable to the massive influx of 'undocumented' or illegal immigrants, principally from Mexico, whose insecure residential status in the USA has provided local employers with 'perhaps the largest pool of cheap, manipulable, and easily dischargeable labor of any advanced capitalist city' (Soja,

Morales & Wolff 1983, p.219). Elsewhere, it is in the Sunbelt's small cities and non-metropolitan areas where 'legal' low-wage employees do not represent politically significant concentrations, that job-creation in the manufacturing sector has been most noticeable.

National differences in political economy and culture are evident in the experience of inner-city decline, as they were in periods of growth. Yet the widespread outbreak of rioting in British cities in 1981 forced many people to recognize that problems generally assumed to be peculiar to the USA are in fact more pervasive. Racial discrimination fuelled many of the conflicts, and they were symptomatic of a broadening experience of inner-city deprivation, unemployment, and environmental decay. Perversely, reduced revenue support from the national government has weakened the ability of British cities to respond to these problems (Hausner 1987). Compared to the UK and the USA, Canada has a very limited inner-city problem, although declining manufacturing employment in the core of the three largest cities (Toronto, Montreal, and Vancouver) has been recognized as a trend which planning policies should not needlessly accelerate.

Urban prospects
The postwar growth of cities and metropolitan regions, both physically and in terms of the institutionalized services provided within them (the infrastructure of the welfare state), has been characterized by Harvey (1985a, pp.209 & 211) as 'demand-side urbanization ... in which questions of production and fundamental class relations were held in abeyance.' Preservation of the quality of urban space as an environment for living became a leading issue in urban politics, not only in archetypal 'post-industrial' cities, such as Vancouver, but also in 19th-century industrial cities, such as Montreal (Ley 1983). However, the erosion of industrial employment and municipal finances since the mid-1970s has restored the economic performance of metropolitan areas firmly to its position at the top of the political agenda. Harvey (1985a, pp.212–18) suggests that in the contemporary era of 'supply-side' urbanization, urban regions compete to attract resources with one or more of the following strategies:

(a) By improving their competitive position within the international division of labour. This may involve 'creating a "favourable business climate", as well as corporate handouts and other forms of subsidy to industry.' Spending priorities shift away from projects enhancing 'equity and social justice' toward those promoting 'efficiency, innovation, and rising real rates of [labour] exploitation.'

(b) By improving their appeal as centres of consumption. 'There is more to this than the redistributions achieved through tourism, important and extensive though these may be.' Investment is channelled into the lifestyle infrastructures that attract high-income residents, such as cultural centres, sports stadia, and distinctive (often "heritage") environments, both residential and leisure-oriented. Such a strategy tends to promote 'the public subsidy of consumption by the rich at the expense of local support of the social wage of the poor.'

(c) By competing 'for those key control and command functions in high finance and government that tend, by their very nature, to be highly centralised while embodying immense power.' Nodality within the global or national economy and efficient transportation and communications infrastructure are critical to the success of this strategy. At the peak of the international hierarchy of such centres are the 'world cities' (see below), which are also major centres of consumption.

(d) By competing 'for direct redistribution of economic power'. The networks through which resources flow are largely, although not exclusively, in the hands of higher levels of government. The location of spending on the facilities and payrolls of the welfare state (hospitals, universities), the armed forces, state-owned industries, or private sector firms favoured with government grants or contracts, is the subject of intense political activity, particularly in peripheral regions for which this strategy of economic survival appears the only live option (see Weller 1977).

None of these options is 'costless or free of serious political and economic pitfalls' (Harvey 1985a, p.213). Since the mid-1970s, Los Angeles has prospered on all *four* fronts (with its low-wage manufacturing economy based on undocumented immigrants), but few cities have been so fortunate: 'Cities like Baltimore, Lille and Liverpool, in contrast, [have] scored low on most or all of them with the most dismal of results' (Harvey 1985a, p.219). The heightened instability of the international economy and of corporate prosperity within it has made it difficult to generalize about the experience of cities in Western core nations. Nevertheless, the prospects currently facing specific cities are affected by the age, size, location, and functional specialization of each one.

Control of the non-socialist world's production system is institutionally concentrated within a relatively small number of TNCs and comparable public-sector enterprises. The cities from which these institutions deploy their resources are those whose functional importance is greatest and most durable. At the apex of the international hierarchy of economic control is a set of 'world cities' (Friedmann & Wolff 1982), the pre-eminent characteristic of which is their articulation of the capitalist world's financial markets (Reed 1983). Most core nations contain one such city which, at the national scale, exercises decisive influence

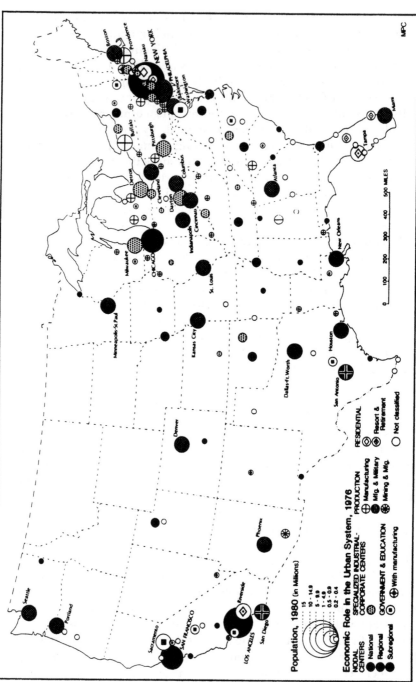

Figure 8.7 Economic rôle of metropolitan areas in the American urban system, 1980 (after Stanback, T. M. *et al.* 1981).
Source: Conzen, M. P. 1983. American cities in profound transition: the new geography of the 1980s. *Journal of Geography* 82; Figure 2.

over economic activity and is the 'gateway' through which interactions between the domestic and international economy are channelled. It is in keeping with the size of the US economy that, in 1980, it contained three of the top ten financial centres (New York, Chicago, and San Francisco). Clearly, metropolitan areas at this level of the international hierarchy are not immune from the economic, demographic, and fiscal problems which have been outlined above. Almost by definition, 'world cities' are very large, and became so before 1945. The troubled state of the world economy since 1973 has increased, rather than decreased, the strategic importance of the financial and specialized producer services on which their urban economy is based, but it has speeded the decline of their previous goods-transforming and -handling specializations.

Stanback et al. (1981) devised a typology of the 140 largest metropolitan areas (SMSAs) in the USA, which indicates the nature of functional specialization within that nation's urban system (Fig. 8.7). Their classi- fication consists of five major groups of cities, identified as *nodal* centres, *functional nodal* centres, *government and education* places, *production* centres, and *residential* centres. *Nodal* centres 'are characterized by economic bases both strongly specialized and diversified in the distributive services and in the complex of corporate [control] activities' (Stanback et al. 1981, p.99). Nodal centres are present in all city-size categories. At the peak of the hierarchy, 'national nodal' centres embody the characteristics of 'world cities'. 'Regional nodal' centres (with populations of more than one million in 1976) 'are somewhat less sophisticated in the range of producer services they offer' (ibid., p.100) but provide for the comprehensive needs of corporate administration in the main regional markets of the country. 'Subregional' centres (SMSAs with populations of 250 000 to one million) have a less diversified service sector, but are important distribution centres and 'often remain closely tied to agricultural markets' (ibid., p.100).

The group identified as *functional nodal* centres consists of 'a somewhat peculiar breed' of cities (ibid., p.100), major manufacturing centres, often containing significant concentrations of corporate administration and R&D activity, but relatively weakly involved in supplying regional con- sumer markets. Detroit and Pittsburgh are the largest cities in this group, and smaller centres include Rochester, New York and Greensboro, North Carolina. (There is a fair degree of conceptual similarity between this group and Malecki's (1980) category of 27 'manufacturing' R&D locations (see p.167), although Malecki's SMSAs in fact divide evenly among the nodal, functional nodal, and manufacturing (production) cat- egories defined here.) Overall, the composition of Noyelle & Stanback's nodal and functional nodal groups confirms Pred's (1977) argument that economic control within most contemporary urban systems is quite noticeably diffused through the size-hierarchy of cities, but it equally stresses the strategic advantage derived from the agglomeration of diversified financial and corporate services still enjoyed by most of the larger metropolises.

A group of cities varying in size from Washington, D.C. to Ann Arbor, Michigan, but predominantly in the 0.25–0.5 million size range, is identified as *government and education* places by an 'overconcentration of employment in the government and/or nonprofit sectors' (Stanback *et al.*, 1981, p.101). These urban areas have tended to grow rapidly in parallel with the rising volume of public-sector spending and employment and, although much of that stimulus has abated, they have generally escaped the problems of job loss associated with recession in the manufacturing sector. *Residential* centres are another group of SMSAs which have continued to experience rapid growth, either as favoured suburban metropolises of national nodal centres, or as resort or retirement centres with a warm winter climate. Noyelle & Stanback's fifth group, *production* centres, subdivided into 'manufacturing', 'industrial and military' and 'mining and manufacturing' categories, is predominantly made up of cities which are smaller than the *functional nodal* centres, but which are distinguished less by size than by the relative absence of control functions (planning, administration, or R&D) attached to their production enterprises.

The map of relative prosperity in core nations such as the USA has become increasingly marked by *intraregional* variation. This may be taken as evidence to support the claim that a city's functional specialization within the national and international economy is likely to be the principal indicator of its state of economic health. Other variables are secondary, but certainly not insignificant. In the US context, the differential performance of the traditional manufacturing core and the southern and western periphery has naturally been reflected in the differential growth rates of their urban areas. The fact that smaller SMSAs have, in recent years, been growing faster than large ones reflects the complex interaction of age, size, and location, for the older and larger metropolitan areas, the central-city problems of which have been identified above, are disproportionately concentrated in the traditional Manufacturing Belt.

The contemporary economic prospects of British cities can be analysed in terms of the same variables. One of the greatest obstacles preventing the economic rejuvenation of British regions is the pronounced polarization of control functions and of the supporting network of specialized producer services in London and the towns of its commutershed. Manchester, Glasgow, and Birmingham are not in the same league as Chicago or Boston as regional foci of financial and corporate administration. In 1971, Central London alone contained four times the *combined* total of professional and administrative workers in

Britain's other conurbation centres (Daniels 1977). Takeover activity in
the UK results almost without exception in a loss of decision-making
by peripheral regions to the South East (McEnery 1981). As in the
USA, the age of urban centres is significant, especially in differentiating
those which developed a relatively narrow industrial base during the
19th century from those whose manufacturing employment is based
on more modern technologies (although the recession of the 1980s
has severely affected many of the latter centres, such as Coventry,
also). The locational component of British urban growth differentials
is closely linked to the age variable, for it is the regions of 19th-
century industrialization which are the major sources of interregional
outmigration, whereas population growth has been most noticeable in
the towns of the relatively rural regions of East Anglia and the South
West (Regional Studies Association 1983). Indeed, the regional pattern
of prosperity in England & Wales in the late 1980s is much closer to that
of 1700 than to that of 1800 (see Fig. 4.2).

The non-metropolitan economy

Changing global markets and the escalating costs of traditional public
policies have, since 1980, begun to have major consequences for the
agricultural sector in Western industrialized nations. Throughout the
period since 1945, as the farm population has shrunk steadily in all
states where it was not already (as in the UK) under 5% of the
labour force, governments have provided a wide range of financial
support to agricultural producers. Strong cultural values and political
considerations, different in origin and detail in North America as
compared to Western Europe but similar in focus and effect, have
prompted a public commitment to 'the family farm'. Simultaneously,
however, strong economic pressures (the 'cost–price squeeze') have forced
farmers into a treadmill of expanding their holdings and increasing the
capital-intensity of their production simply to maintain their incomes
in the face of shrinking profit margins and the increasingly stringent
demands of food processors (Vogeler 1981, Gregor 1982). In North
America and the UK particularly, farmland has become steadily
concentrated into fewer, larger holdings.

The cost of agricultural support, regarded in the past as a relatively
modest premium on a policy with many intangible benefits, became
more controversial in the 1970s and more burdensome in the mid-1980s.
Given the bias of assistance towards large farmers (with entitlements
frequently based on volume of production), the rapid rise in grain prices

in the mid-1970s was widely seen as subsidizing windfall profits for the already affluent. In addition, public sympathy towards farmers in Britain was eroded as the aesthetic consequences of untempered profit-maximizing arable production (hedgerow removal, wetland drainage, etc.) began to emerge (Bowers & Cheshire 1983). (The environmental consequences were discussed in Ch. 3.) Nevertheless, despite growing opposition, agricultural subsidies have continued to increase in both the USA and the EEC (and also in Japan).

The 'crisis' which has overtaken the North American farm economy since the early 1980s and which is impending in Western Europe is the result of simultaneous drops in agricultural commodity prices and in the capacity of governments to support farm incomes. The Common Agricultural Policy (CAP) of the EEC, which is geared to provide an adequate income to the least efficient small farms in France and West Germany (which are a legacy of the land fragmentation of much of former 'common-field Europe'; see Ch. 4), has become increasingly out of touch with declining trends in 'world' agricultural prices (World Bank 1986). In 1986, the direct subsidy costs of the CAP were $23 billion, and this cost was doubled by the inflated food prices which consumers paid. Subsidized sales of surplus produce, particularly of grain (of which the EEC was a substantial importer in the 1960s), have aggravated the financial difficulties of many North American farmers. The cost of agricultural subsidies to address these problems in the USA rose from $3 billion in 1981 to $30 billion in 1986 (15% of the federal deficit; *Financial Times* 18 December 1986).

Trends in the US farm economy, which are evident in a muted form in Canada, the UK, and Australia, are rapidly polarizing production units and destroying traditional images of 'family farming'. Food production in core nations has become the activity of an integrated *agribusiness system* (Wallace 1985), in which the major actors are oligopolistic food retailing and manufacturing firms, farm input suppliers, and a small group of large, well capitalized farms which produce a major share of marketed agricultural output (and would produce a much greater share if subsidies were reduced). In 1982, 28000 farms (1.2% of the total), with sales of over $500000 each, generated 33% of gross farm revenue in the USA and 64% of net farm income. At the other end of the spectrum, 1.36 million farms (61% of the total) generated 6% of gross revenue and had aggregate financial losses equivalent to −4% of net farm receipts. These producers are fundamentally part-timers, dependent on off-farm sources of income (US Congress 1986). Farms which most closely match the traditional image, are the 'shrinking middle' of the agricultural sector. These are the ones which have been most severely hit since 1980 by heavy debt, declining equity (as land values have dropped, by as much as 50% in some areas) and severely reduced cash flow. Farm bankruptcies and the

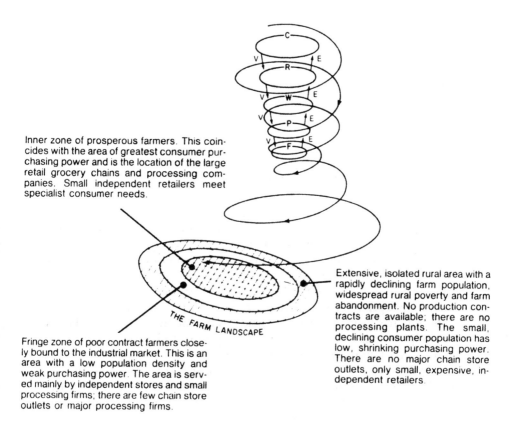

Inner zone of prosperous farmers. This coincides with the area of greatest consumer purchasing power and is the location of the large retail grocery chains and processing companies. Small independent retailers meet specialist consumer needs.

Extensive, isolated rural area with a rapidly declining farm population, widespread rural poverty and farm abandonment. No production contracts are available; there are no processing plants. The small, declining consumer population has low, shrinking purchasing power. There are no major chain store outlets, only small, expensive, independent retailers.

Fringe zone of poor contract farmers closely bound to the industrial market. This is an area with a low population density and weak purchasing power. The area is served mainly by independent stores and small processing firms; there are few chain store outlets or major processing firms.

Figure 8.8 Schematic zonation of the agribusiness system focused on a large metropolitan market.
Source: Smith, W. 1984. The 'vortex model' and the changing agricultural landscape of Quebec. *Canadian Geographer* 28; Figure 2.

insolvency of rural banks are resulting in dramatic changes to the rural economy of the US Mid-West.

The emerging configuration of the agricultural economic landscape around major metropolitan regions of core nations in response to the pressures within the agribusiness system (Smith 1984) is suggested in Figure 8.8. The 'industrialization' of agriculture (Troughton 1982) is thus one of many factors undermining the traditional distinction between 'rural' and 'urban' areas in Western industrialized nations (Pacione 1984). The deconcentration of population and manufacturing employment away from large metropolitan centres has been noted above. Rising discretional incomes and improved highway networks have greatly increased the use of rural areas for recreation and seasonal (second) home ownership. The nature and the areal extent of these activities varies between 'crowded' Europe and the northeastern USA and 'empty' regions in western North America and Australia, and between cultures that appreciate 'the countryside' and those that favour 'the wilderness', but the penetration of the metropolitan economy into areas that were formerly dependent on primary-sector employment is pervasive. This still leaves many remote peripheral regions with very limited economic prospects.

The politics of prospective geographies

The apparent failure of policies of macro-economic management to shield the mixed-capitalist core nations from the economic dislocations of the 1970s and 1980s (rising inflation and unemployment, declining productivity growth, growing public (and private) indebtedness, etc.) provided a political climate in which the value of the welfare state and the propriety of pervasive government intervention in the national economy has been decisively questioned for the first time since the start of the post-1945 boom. The dominant political philosophies in the US and in the UK in the 1980s embodied the belief that the State has become an intolerable burden on the economic process of wealth creation and on the freedom of individuals to make responsible choices to shape their own lives. Underlying the political rhetoric is a recognition that national economies which are struggling to adjust to a much more competitive global economy than that of the 1960s cannot, *other things being equal* (and therein lies the set of political choices), sustain the levels of public consumption and income redistribution which had come to be regarded as the norm by 1970. The thrust of domestic policies has therefore been towards a relative weakening of public provision for those who are less well off, and a relative fall in the public appropriation (through taxation) of the income of the more prosperous individuals and corporations. The decline in regional policy expenditures is a very minor, but spatially explicit, symptom of this trend. The favoured priorities in the current environment are those designed to foster initiative, entrepreneurship, and self-reliance.

For a variety of reasons, the prosperity of individual members (or would-be members) of the workforce, and of individual places, is becoming more dichotomized. What is behind this increasing polarization of wellbeing? It was argued in Chapter 2 that once a society bursts the bounds of local self-sufficiency the social stratification of wealth accumulation becomes indivisibly associated with its geographical realization. In Chapter 6 the concentration of productive activity within the capitalist world-economy was identified as being in the hands of a relatively small number of corporations which, in the past few decades, have increasingly organized themselves to benefit from specific forms of the division of labour at a global scale. Technological developments have tended to dichotomize the production process into tasks which require high-level technical and managerial skills and those which require relatively unskilled manual labour. The pace of technological advance has been accelerated by competitive pressures to reduce production costs, and has been achieved disproportionately at the expense of high-wage skilled labour. Simultaneously with the polarization of skill requirements and financial rewards among the labour force has come a greater geographical polarization (itself made possible by developments in transport and communications technologies) of the workplaces within which these contrasted forms of production are carried out. We can see, in this context, why the economic prosperity of specific urban centres has become more closely tied than in the past to their particular functional rôle within a national or global network of corporate production operations, and more loosely tied to the general level of economic activity in the region in which they are located.

To set against these processes of polarization, it was argued in Chapter 5 that the essential feature of the liberal–democratic, mixed-capitalist nations is that the State survives by accommodating a plurality of conflicting interests and is, indeed, legitimated by its success in managing or containing tendencies toward social polarization. Pronounced socio-economic cleavages and persistent regional economic disparities are a threat to that legitimacy, over and above any more narrowly defined economic handicaps they may represent to national wellbeing. The evolution of the welfare state has exerted a powerful check on the tendencies towards socio-economic polarization intrinsic in market-based societies.

It is a symptom of the erosion of the welfare state that the marked contrasts in living standards and economic security which Marx observed in the society of early-19th-century England, and which characterize the majority of Third World states today (see Ch. 11), are tending to reappear. Granted that governments claim to be paring down the welfare state rather than abolishing it altogether, core States are increasingly unable (because of large fiscal deficits) or unwilling (on grounds of political philosophy) to protect, let alone improve, the wellbeing of a large number of their citizens. Those whose employment prospects are severely restricted by their lack of qualifications for the 'good' jobs within a high-technology and information-based economy, or whose income depends substantially on transfer payments (notably the unemployed and a majority of senior citizens), are the most vulnerable. They are concentrated in inner cities and old industrial towns and, without some radical changes in the political and economic environment, their prospects are not bright (Berry 1980, Eversley 1980). In the ghettoes of New York, growing numbers of AIDS victims are the latest manifestation of the economic malaise which gives rise to a persistently pathological urban society (Norton 1987).

9 Industrialized state-socialist economies

Nature and origins

Only one distinctively different economic system has emerged since the capitalist world-economy reached fully global proportions in the early 20th century. Pioneered by the USSR, it consists of those *state-socialist* economies in which there is all-encompassing political control of the national economy by the State bureaucracy. Yet writing in 1988, one must immediately sound a note of caution about the tenses in which the system is described, for the 1980s have witnessed potentially revolutionary developments in many socialist states, not least in the USSR (and in China, which is not covered in this chapter). It is premature to speculate here whether the economic and political shortcomings of state-socialism, which are clearly perceived by the reformers who are attempting to reshape it, will eventually lead to its demise. The increased rôle of markets as mechanisms of exchange and economic coordination is central to most of the changes currently taking place. The extent to which this leads to substantial reintegration of state-socialist economies into the capitalist world-economy will not be determined for many years yet. By their very nature, Soviet-style economies do not shed well established characteristics overnight, however vigorously restructuring (*perestroika*) is pursued.

Marx was an astute and detailed critic of the economy of industrial capitalism which he observed taking shape in England during the middle of the 19th century. He expected the social conflict and the contradictory developments embedded within the capitalist system to give birth, inevitably, to a socialist society, but only when the technical successes of industrial capitalism had had their full impact. What Marx did *not* provide was a similarly perceptive and detailed analysis of how a functioning socialist society might appear. His utopian sketches of the nature of personal freedom in a world which had ceased to experience exploitation and scarcity are far removed from the sort of down-to-earth insights which a newly established socialist regime might welcome as a basis for policy-making. One therefore has to distinguish Marxism as a philosophy, ideology, or mode of historical analysis from the actual substance and behaviour of socialist societies which derive their existence from the application of Marxist thought.

It is a fact of history that the USSR was the first socialist state, and that it was Russian Marxists who had to grapple for the first time ever with the practical problems of making a socialist economy work. This specifically Russian context, which includes the nation's cultural and economic history and its geographical characteristics, cannot be divorced from an analysis of the development of the first industrialized society to be organized on Marxist principles. Because the Eastern European states, which were brought into the communist bloc following the Second World War, had their economies reorganized more or less on Soviet lines, there has been essentially only one form of industrialized socialist society to compare with industrialized capitalist economies. Alternative and, to many people, preferable expressions of Marxian social organization within a highly developed economy have been attempted (notably in Hungary, Poland, and Czechoslovakia), but prior to the mid-1980s such initiatives were seen by the USSR as a threat to its interests, and were suppressed. The currently more tolerant climate towards reform in the USSR appears to give more scope for Eastern European socialism to assume diverse national forms, much as the capitalist economies of Western Europe differ among themselves in significant respects.

The task facing the Bolshevik revolutionaries when they seized power in the chaotic conditions of 1917 was to adapt Marxism 'to the political and economic situation of a relatively backward country' (Nove 1969, p.33). Although the level of economic development in Russia had differed little from that of all but the most advanced areas of Europe in the middle of the 18th century, by the middle of the 19th century the weakness of the Russian state in a world of industrializing nations was made inescapably clear. A landed aristocracy continued to rule over an essentially feudal society, within which the peasantry had been reduced to serfdom. Trade and manufacturing were poorly developed, for want of effective demand, and so Russia lacked the growing class of merchants and entrepreneurs whose economic power and political influence had long been transforming society in the core areas of Western Europe. Lack of a modernized transportation system was a major factor in Russia's defeat by British and French forces, operating 2000km from home, in the Crimean War (1854–6). The shock of this defeat stimulated a series of reforms by which the aristocracy of the Tsarist regime sought to preserve its place in the world without compromising its status at home. However, attempts at rapid economic development quickly came up against the contradictions inherent in this ambiguity and the bottlenecks created by past conservatism (Kemp 1969).

Serfdom was abolished in 1861, but much of the old institutional structure of agrarian society remained, so poverty and low productivity remained the dominant characteristics of the rural masses and peasant unrest grew. In the absence of an entrepreneurial middle class and of private capital to finance industrial expansion, the State had to take a leading rôle, notably in railway construction, and encourage foreign investment. Despite considerable achievements by 1914, in a country the size of Russia and among a people so ill prepared in outlook and skills for industrialization, economic development was highly localized, both geographically and sectorally. Modern industry was concentrated in the two great cities of St Petersburg (Leningrad) and Moscow, in 'Russian Poland', and in the Donets Basin (Nove 1969). Although the Trans-Siberian Railway extended across Asia to Vladivostok, economic development had hardly scratched the surface of the Russian Empire outside these regions. Metallurgical, textile, and food-processing industries had been established, but almost no machinery was produced domestically. The only source of export earnings to pay for imports of capital equipment and to service foreign borrowing was the backward agricultural sector. Significantly, it was the privations of an uncertain and inadequate food supply, on top of the wretched living conditions faced by workers in the least healthy and most expensive of all contemporary European capitals, which finally triggered the Bolshevik uprising in St Petersburg in 1917 (Bater 1979).

Not until a decade after the revolution did the distinguishing characteristics of the Soviet economic system take shape. The novel challenges of establishing a workable socialist society against a background of international conflict, civil war, and the problems of adjusting to the loss of territory which had contained a disproportionate share of pre-1914 manufacturing capacity involved much initial pragmatism (Nove 1969). By the mid-1920s, however, the fundamental considerations governing the Soviet Union's development ambitions were clear, and so was the nature of the difficulties which the communist regime would encounter. The solutions adopted under Stalin achieved 'success', but at a high price to the individuals who perished in the process and to subsequent national development because of the structural inflexibility of the economic system which he bequeathed.

A state-socialist economy of the Soviet variety came to be characterized by:

(1) *state ownership* of the means of production, in which context the treatment of agricultural land has given rise to the greatest problems;

(2) *command economy*, in which directive planning replaces market mechanisms as the basis for allocating resources and specifying outputs;
(3) *centralized control*, so that all major decisions are made by State agencies in the capital and filtered down through a myriad of hierarchical channels to workplaces throughout the country.

Within this form of economic organization, political and economic decision-making are much more closely intertwined than they are in non-communist states. In the absence of market signals, price levels in different sectors of the economy can be set to further politically desirable objectives, and investment can be channelled to reflect the regime's priorities rather than those dictated by comparative profitability in a market economy.

If the principal objective is 'to build communism' by developing the country's productive forces along socialist lines, then the central organs of administration are theoretically in a position to harness the entire economic system to achieve this desired end. In practice, one of the greatest problems of a centralized, state-socialist economy is the difficulty of translating a directive plan for the economy as a whole, or some sector of it, into a workable operational plan to guide the daily activity of individual production establishments. Marx regarded the decentralized and uncoordinated planning of capitalist enterprises as anarchic, but he never seriously considered the constraints which render a theoretically more rational nationwide plan made by a centralized authority almost impossible to administer without generating counterproductive responses (Nove 1977).

Before analyzing the economic geography of the industrialized state-socialist nations, it is worth returning to the question of their 'Russian-ness'. Unlike many of its European neighbours, Russia had never espoused Western liberal-democratic traditions, so the centralized authority of the Soviet regime represents no decisive break from the autocracy of the Tsars. A continuing source of friction in Eastern Europe since the death of Stalin has been Soviet unease with, and ultimate intolerance of, moves towards a more open and flexible socialist socio-economic system by nations where illiberal, authoritarian administration encounters much stronger cultural opposition. Many Eastern European writers have argued for the desirability of a regulated market system within a socialist state, both on grounds of economic efficiency and of its contribution to genuine political freedom (Selucky 1979), but this implies a devolution of control from the centre which has been (prior to Gorbachev) an anathema to Soviet thinking. In the industrial sphere, the relative underdevelopment and limited market of early 20th century Russia led to the formation of state-sponsored cartels in sectors such as coal mining and metal production, which reinforced the pattern of centralized administration (Kemp 1969).

The priorities and strategies of the Soviet régime have also been influenced by more explicitly geographical factors. Invasions from the west under Napoleon and Hitler highlight the historic threat under which Russians have felt compelled, since the end of the Second World War, to ensure the existence of subservient states in Eastern Europe. Retreat into the country's vast continental landmass has been the classic Russian response to these invasions, and this defence consideration played a prominent rôle in the geographical pattern of industrial development during the early Soviet era. Although the USSR's natural resource endowment is vast, so too are the distances separating many of these resources from their natural markets, and environmental conditions frequently constrain their exploitation. Comparisons between the functioning of state-socialist and capitalist economic systems need, therefore, to be made against the background of the specific conditions facing Soviet producers, as well as of those facing, say, their North American counterparts. Nowhere is this more necessary than in the agricultural sector.

The nature and performance of Soviet agriculture

The leadership of the Soviet communist party made a clear-cut decision in the late 1920s to embark on a programme of rapid industrialization at the expense of agricultural development and of the living standards of agricultural workers. The first Five-Year Plan (1928–32) concentrated on the construction of heavy industry, and the rôle of the agricultural sector was defined as providing food for a rapidly growing urban labour force, and providing the economic surplus which would finance the ambitious capital investment targets of the plan. Herein was a contradiction, for extra marketable output was required from the peasant agricultural producers to feed industrial workers, yet the terms of trade between agricultural and industrial goods were to be set very unfavourably to the peasants, giving them no incentive to expand their production in keeping with the regime's priorities (a blunder repeated more recently in some Third World states [see Ch. 11]). Stalin's solution to this problem was to push through a rapid collectivization of peasant properties, enforcing co-operative cultivation of the larger farm units which resulted, from which compulsory deliveries of grain were made to the State procurement agencies. This sudden move, a 'revolution from above', met with widespread peasant opposition which took the form, in particular, of a mass slaughter of livestock (Nove 1969).

In 1928, individual peasant holdings accounted for over 97% of the area sown to crops in the USSR. State-owned farms covered only 1.5%, and various forms of collective farms only 1.2%. By 1930, one-third of the crop area and almost a quarter of peasant households had been collectivized; by 1931 two-thirds of the area and over half the households had been reorganized on this basis; and by 1935 the totals had reached 94% of the crop area and 83% of peasant households. Such a rapid and radical revolution in the structure of an agrarian economy is without parallel: it involved severe losses and dislocation to the rural economy and brought mass starvation to some regions. Agricultural production as a whole took a decade to recover its 1928 levels of output, and meat products did not do so until 1953 (Fig. 9.1). But a steady rise in crop production represented the success of the collectivization programme from Stalin's point of view. Grain and industrial crops were made available to the workers and industries of the USSR's rapidly growing cities. Recovery from the initial losses of the collectivization drive would obviously have been faster if the devastation of the Second World War had not intervened.

The institutional structure of Soviet agriculture has remained relatively stable since Stalin's death (1953). The two principal forms of farm operation are the state farm (sovkhoz) and the collective farm (kolkhoz). Initially, state farms were established primarily in peripheral areas without a long agricultural tradition, whereas collective farms were concentrated in the existing major regions of agrarian production, especially the grain lands of the Ukraine and lower Volga (Fig. 9.2). By 1976 the proportion of arable land contained within state farms had risen to 53%, having expanded rapidly in the late 1950s as a result of the Virgin Lands campaign (see below), and having maintained a slow increase subsequently through the conversion of kolkhozy into sovkhozy. During this time the operational and administrative distinction between the two sorts of enterprise was greatly reduced. Until 1958, collective farms were operated in conjunction with State-controlled machine tractor stations (MTS), which served the dual purpose of providing machinery to the kolkhozy and ensuring the political and economic compliance of their workers with State policies. The equipment owned by the MTS was subsequently sold to the kolkhozy, which were assumed capable at that stage of maintaining it themselves.

Families living on collective farms were, from the beginning, allowed to maintain a small private plot. This was seen by the regime as a temporary concession to a peasantry that was not yet ready to embrace a fully socialized economy. These plots have eventually achieved socialist legitimation, however, which is not surprising in view of their crucial contribution to aggregate agricultural production. Rights to a plot have been extended to workers on state farms and (where land is available) to

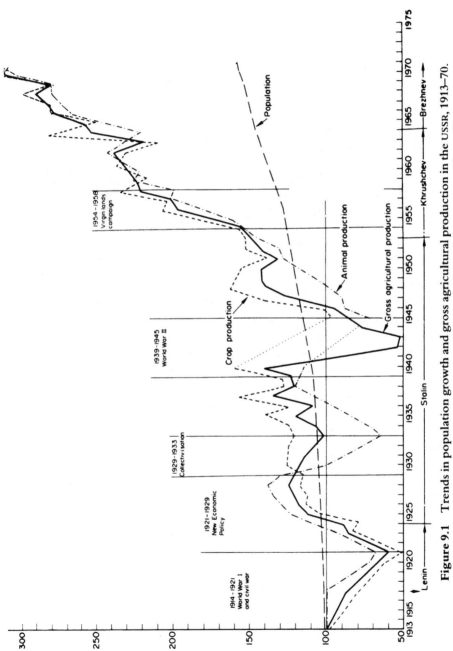

Figure 9.1 Trends in population growth and gross agricultural production in the USSR, 1913–70.
Source: Bergmann, T. 1975. *Farm policies in socialist countries.* Farnborough, Hants: Saxon House; Figure 2.8.

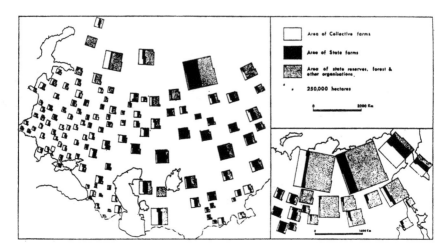

Figure 9.2   Distribution of agricultural land in the USSR, by sector, 1965.
Source: Symons, L. 1972. *Russian Agriculture: a geographical survey*. London: Bell; Figure 1.

suburban factory and office workers. Although occupying only 4% of all arable land, private plots provide 27% of gross agricultural output, concentrated in labour-intensive sectors such as livestock products and fresh fruit and vegetable production (Lydolph 1979).

Soviet agriculture has recorded impressive increases in production since the Second World War, but in the all-important grain sector these have not been accompanied by a comparable reduction in year-to-year fluctuations (Fig. 9.3). This is one indication of the degree to which much of the USSR experiences climatic conditions which are adverse to arable farming, and which create semi-marginal regions vulnerable to annual changes in weather patterns. Field's (1968) attempt to quantify the most significant climatic determinants of agricultural productivity, and to provide a basis for comparison with North American conditions, makes clear the handicap from which Soviet farmers suffer. He classified arable land in terms of a thermal index (summer degree-months above 0°C) and a water balance index (actual as a percentage of potential evapotranspiration). His data (Table 9.1) indicate that the least favourable categories of each index, added together, incorporate 93% of cropland in the USSR, but only 34% of cropland in the USA. At the other end of the scale, land which enjoys the best of both moisture and temperature conditions represents 19% of the US total, but only 0.3% of the Soviet total. In other words, it is clear that disparaging comparisons of Soviet agricultural productivity with that of the USA need considerable qualification.

Table 9.1 Cropland of the USSR and of North America classified by thermal and moisture regimes (percentage distributions).

(AE/PE) × 100 (%)	Degree-months, Canada				Degree-months, USA				Degree-months, Canada & USA				Degree-months, USSR			
	100–199	200–299	300+	total	100–199	200–299	300+	total	100–199	200–299	300+	total	100–199	200–299	300+	total
90–100	22	2		24	8	30	19	57	10	25	15	50	26	0.1	0.3	26
80–89	10			10	2	5	2	9	4	4	2	10	14	1		15
65–79	48			48	5	7	3	15	13	5	3	21	18	6		24
0–64	18			18	4	7	8	19	7	5	7	19	22	9	4	35
totals	98	2		100	19	49	32	100	34	39	27	100	80	16	4	100

AE = actual evapotranspiration; PE = potential evapotranspiration.
Source: Field, N. C. 1968. *Environmental quality and land productivity: a comparison of the agricultural land base of the USSR and North America,* Canadian Geographer **12**; Table V.

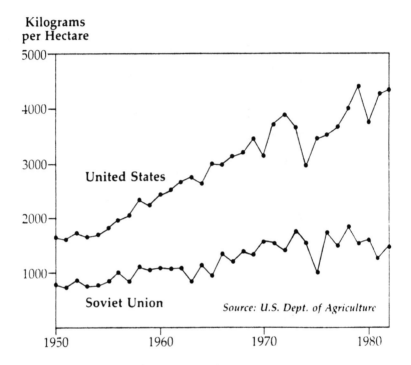

Figure 9.3 Soviet and US grain production, 1950–82 (United States Department of Agriculture data).
Source: Brown, L. R. 1982. *U.S. and Soviet agriculture: The shifting balance of power.* Worldwatch Paper 52, Washington: Worldwatch Institute, p.8.

The socialist organization of agricultural production in the USSR creates a different set of geographical distributions than would prevail were it an advanced capitalist economy. National self-sufficiency is an important goal in agriculture, as in other sectors of production. As a result, 'crops with limited environmental tolerances are given priority in areas where they *can* grow ... even though they might not yield the highest returns' (Lydolph 1979, p.235). Cotton, tea, and citrus fruits are among the crops whose production is fostered by this policy. At the national level of planning, the second priority in deciding regional specializations goes to crops giving the highest yield, allowing corn, for instance, to displace wheat in the southeastern Ukraine. In the absence of these priority demands, crop production in a given region tends to reflect that best suited to the environmental conditions. The environmentalism implied in this ranking, which to some extent accords with natural comparative advantage, is modified by other considerations. Increasingly, specialization in perishable dairy and market-garden produce has been encouraged around major urban centres. Conversely, in agriculture as in other sectors of the economy, an ideological and pragmatic desire to minimize transportation outlays leads to greater regional self-sufficiency in food production than would be encouraged in a market economy. This diversification has even been imposed at the farm level, with a resultant loss in productivity (Nove 1977).

Because Marxist orthodoxy claims that land rent is a capitalist phenomenon which is abolished by the socialist revolution, Soviet planners have not found it easy to adjust to the realities of locational advantage (location rent) in setting prices for agricultural products. Prior to the reforms of 1958, there was no geographical variation in state procurement prices for any particular crop. Given the diversity of environmental conditions across a nation the size of the USSR, this would have produced marked disparities in farm revenue across the country had there not been regional variations in the level of compulsory crop deliveries, and in the amount of payment in kind levied on the *kolkhozy* for MTS work (Nove 1977). By 1958, an explicitly regionalized set of procurement prices was deemed ideologically acceptable, justified in terms of spatial equity and as an encouragement to appropriate regional specialization. Jensen's (1969) analysis of 1958 wheat price zones notes, however, that they were too large to meet their objectives and, moreover, they followed administrative rather than 'natural-economic' boundaries, so creating some irrationally sharp price gradients. Nor did the overall pattern of price differentials adequately reflect regional differences in production costs. Even when allowance is made for subsequent increases in price differentials and in the number of zones, official recognition of locational advantage within Soviet agriculture remains half-hearted. The systematic evaluation of the potential of agricultural land was long rejected on ideological grounds, and the use of its findings to delineate a more refined system of price zones would almost certainly undermine the power structure of the existing administrative regions because boundaries would be unlikely to coincide (Jensen 1969).

The substantial increase in Soviet crop production since the early 1950s has been achieved through a combination of cultivating existing land more intensively and bringing new land into production. Grain output rose from 81 million tonnes in 1950 to a peak of 229 million in 1978 (although the average harvest in the mid-1980s was 20% below this). Yields almost doubled over this period, with the help of a ten-fold increase in chemical fertilizer production and a more than three-fold increase in the number of tractors deployed (Bergmann 1975). The most dramatic development of the period, however, was the expansion of grain cultivation in the 'Virgin Lands' of northern Kazakhstan and adjacent regions. Between 1953 and 1958 the area of cropland in the USSR increased by 38 million ha, of which 85% was in the Virgin Lands. Despite lower yields in these dry marginal regions, total Soviet wheat production rose from 41.3 to 76.6 million tonnes, with the share of the Virgin Lands rising from 38 to 57%. This expansion of wheat acreage allowed fodder crops, notably corn, to occupy better wheatland elsewhere in support of the regime's priority goal of increasing livestock production (Lydolph 1979).

By the mid-1970s, when the agricultural area of the Virgin Lands had shrunken somewhat from the territory occupied in the enthusiasm of initial settlement, Soviet attention turned away from what has been called 'the drive for freely bestowed fertility' (the extensive cultivation of new land) to the potential for increased production by intensified cultivation of relatively neglected territory. Plans for the Nonchernozem Zone of European Russia to double its grain output between 1976 and 1991 depended on substantial investment in rural infrastructure (land drainage, road construction, etc.) if investment in farm equipment were to be effectively used (Rostankowski 1980). Among the benefits of additional grain production in this populated core region, beyond the further support of livestock production, will be a reduction in the need for long-distance grain shipments from peripheral regions. Symons (1972) notes that between 1950 and 1960, the decade of the Virgin Lands campaign, the mean haul of grain in the Soviet Union rose from 795 to 1152 km.

Inadequate farm-to-market transportation (both roads and vehicles) is one of the principal sources of inefficiency in modern Soviet agriculture. It leads to losses and deterioration of produce and consumes a disproportionate share of farm workers' time (Nove 1977). Low levels of education and technical training among the rural population also continue to be a barrier to productivity improvements. As recently as 1970, half the rural workers had only 0–3 years of primary education, and a further 30% had incomplete secondary education (Abouchar 1979). This was partly a reflection of the abnormal age–sex distribution of the Soviet population which, together with selective outmigration, left a concentration of older women in rural areas. But the unattractiveness of rural life (real and perceived) has made it difficult to hold qualified young people, hampering the attainment of productivity increases commensurate with the proportion of state expenditures devoted to agricultural production. In 1980, for instance, a quarter of the farms in the Soviet Republic had more tractors than they had trained operators (Khinchuk 1987).

Thus, despite its considerable achievements, Soviet agriculture remains a weak link in the national economy, and the USSR has been dependent, against its wishes, on agricultural imports. Grain is the most critical element and it has a high political profile, emphasized by the embargo on increased US grain shipments to the USSR following its 1979 invasion of Afghanistan. Although imports from the capitalist world were sought initially, in the mid-1960s, to make up for weather-related harvest deficiencies in food grains, the USSR has since become a steady purchaser of feed grains to boost its domestic meat production. Internal political pressures require that the Soviet consumer be enabled to approach more closely to North American levels of meat consumption; but, as noted above, the USSR does not possess the environmental basis for the feed-beef economy of the USA. It has no 'Corn Belt'. The best grain land in the Ukraine is comparable to that of the Dakotas, not Kansas. Moreover, in recent years, grain yields have increased much more rapidly in the USA as a result of the more effective management of industrial inputs (fertilizers, etc.).

The disappointing performance of Soviet agriculture in recent years has not been for lack of capital investment; indeed, the sector has been absorbing almost one-third of such funds. The basic problem is that agriculture lends itself to centralized planning least of all economic activities. Productivity could rise dramatically (as the output from private plots indicates) were individual farm units allowed a greater degree of autonomy and given appropriate incentives. This would imply more reliance on market mechanisms than the bureaucrats in Moscow have tolerated in the past, although reforms introduced since 1986 have moved in that direction.

Industrial structure and location

The modern industrial capacity which the Soviet regime inherited from Tsarist Russia was concentrated in a few sectors and a few regions. Geographically, the majority of production took place in the west. When rapid industrialization commenced in 1928, on the basis of the USSR's first Five-Year Plan, the administration set out to redress the spatial imbalance, and to impose its own priorities on the sectoral composition of output. Ideological and strategic reasons combined to favour expansion of manufacturing capacity in the east. The achievement of regional equality in levels of economic activity, in order fully to utilize all human and natural resources of the country, accords with socialist principles. Moreover, Russia's experience of invasions from central Europe pointed to the need to reduce the nations' vulnerability to such attacks by developing an industrial base well into the interior of its continental landmass. From a sectoral point of view, the marked concentration on heavy industry, especially metals, chemicals and engineering, could also be justified from a number of perspectives. Sustained national economic development would be impossible without this basic provision; military capability certainly depended upon it; and Marx had clearly argued for the priority of the capital goods over the consumer goods sector. (A ratio of 3:1 in the capital goods share of industrial output still characterized the Soviet economy in the 1980s [Shabad 1986]).

Symbolic of the entire thrust of Stalin's industrialization programme was the decision to construct the Urals–Kuznetsk metallurgical combine. This involved the complementary development of steel plants 2000 km apart, one close to iron ore reserves in the southern Urals and the other in a rich coal basin in Western Siberia (Lydolph 1979). Nevertheless, for a variety of reasons, the overall pattern of industrial development prior to the Second World War tended to reinforce the inherited bias towards European Russia. The end of the first Five-Year Plan coincided with the seizure of power by Hitler in Germany, which made the likelihood of war much more imminent. The Soviet authorities thus chose to concentrate investment in established industrial regions, where output could be increased more rapidly. In this, their interests coincided with the managers of Moscow– and Leningrad-based enterprises, who had displayed 'hidden resistance . . . against relocation in the primitive regions of the eastern USSR' (Koropeckyj 1967, p.239). As a result, the Germany invasion 'did more to change the distribution of Soviet industrial production than either the pre-war or the post-war plans. Over 1500 industrial enterprises were moved between July and November 1941' (Parker 1968, p.346). Their machinery was dismantled and transported, with their workers, to the Urals and regions further east.

The perennial Soviet debate about industrial location policy at the national scale has reasserted itself since 1945. One line of argument supports continued expansion in the less industrialized east and south--east, where there are many untapped resources, but where extensive investment in basic infrastructure is needed before manufacturing industries can be developed. The other argument is that a more efficient allocation of investment will be achieved if the established manufacturing centres of European Russia, which are also the major market concentrations, are further expanded (Rodgers 1974, Schiffer 1985). The military considerations which were decisive at an earlier time are relatively insignificant today.

Is there a distinctively socialist industrial location theory? The short answer is 'No'. Early Soviet planners found Alfred Weber's (1909) theory appealing: its emphasis on the minimization of transportation costs in the location of weight-losing, resource-based industries was highly appropriate to the sort of investment decisions being made in the late 1920s. (The Russian translation appeared three years before the English was published). During the same period, Soviet applied mathematicians were seeking innovative approaches to economic planning, from whose work both input–output analysis and linear programming were eventually developed (Nove 1979). Stalin's growing imposition of ideological conformity on Soviet intellectuals in the 1930s put a stop to these developments however, and soon 'being labelled a follower of [non-Marxist] Weber meant automatic long-term imprisonment' (Koropeckyj 1967, p.5).

Following the Second World War, a set of locational principles emerged, among which priorities of implementation tended to shift in response to changes in the broader political environment (Barr 1974). In summary, they consisted of:

(1) minimizing the demand for transportation by appropriate location of resource-oriented and market-oriented industries;
(2) distributing production among economic regions so as to allow the more efficient utilization of available resources through specialization, and to facilitate the territorial division of labour;
(3) aiming for regional self-sufficiency in production;
(4) distributing industrial capacity evenly throughout the country;
(5) eliminating differences in employment opportunities between urban and rural areas;
(6) industrializing backward regions, especially those inhabited by national and ethnic minorities, so as to promote their social and economic development;
(7) strengthening national defence capability;
(8) facilitating international specialization of production among socialist nations.

Defence considerations (point 7) are considerably less influential than they were 50 years ago. Minimizing transportation demands (point 1, and implicit in point 3) remains important, however, because of the Marxist concern to limit the growth of non-goods-producing sectors of the economy. Yet implementation of point 2 tends to work against this, as the need for linkages between increasingly specialized manufacturing plants expands with the growing sophistication of the economy. Lydolph (1979) notes, for instance, that the USSR's largest automobile plant, despite being highly integrated, receives inputs from more than 30 suppliers, most of which are over 800 km distant. Points 4 and 6 are complementary although not unproblematic. In the early Soviet period the creation of factory employment in underdeveloped peripheral regions, populated by ethnic minorities, was advocated as a means of instilling socialist consciousness. Greater productivity was to be had, however, from establishments in core regions with access to more skilled labour and better infrastructure. In recent years, the distinction between the core and the ethnic periphery (especially central Asia) has increasingly acquired a new dimension because of diverging demographic trends. Against the background of an increasingly tight labour supply at the national level, the question of whether to concentrate investment in the labour-surplus periphery or to encourage ethnic outmigration into the labour-short industrial core regions has become pressing (Clem 1980). Within the core, substantial rural–urban migration in response to differentials in living standards has reduced point 5 to a memento of Marxist orthodoxy. Insofar as point 8 leads to greater industrial interaction between the USSR and its Eastern European partners in Comecon, its influence tends to reinforce the concentration of manufacturing in European Russia. The priority given in the 1986–90 Five-Year Plan to 'modernization and retooling of existing production capacity' as opposed to investment in new establishments points to a European, rather than a Siberian or Central Asian, regional focus (Shabad 1986, p.1).

The organizational structure of industry has geographical significance in the USSR just as it does in capitalist economies. Perhaps the most important distinction between the socialist and capitalist systems is not the *amount* of economic planning which takes place in each, but the degree to which decision-making is devolved and decentralized. Soviet industry has been traditionally organized into individual sectors, each with its own national (all-Union) ministry. Authority is concentrated at the top of a hierarchical administration which reaches down to the individual enterprise, although various forms of semi-decentralization exist, some of them on a regionalized basis (Barr 1974). From the perspective of the enterprise manager, difficulties can arise in connection with the fulfilment of directives from above – which may be cast in terms maladapted to the immediate production environment – and also in connection with lateral dealings with enterprises in other industrial sectors, in contexts where coordination of activity is hampered by the lack of decision-making authority devolved by the national ministries. As a result, uncertainty-reduction is a major objective of Soviet industrial managers. This is reflected, for instance, in their preference for expansion in existing locations rather than ones which would involve distant linkages and 'by the continued propensity of Soviet firms to integrate backward to protect themselves against supply shortages' (Bergson & Levine 1983, p.427), despite the loss of economies of scale which this involves.

Within the basic resource-transforming sectors, such as the metallurgical and forest products industries, related enterprises have been vertically integrated into a number of multifunctional large firms which promote 'geographically localized production to minimize product transfer among related processing stages ... [or] to use raw materials and recycle component materials to maximize efficiency' (Barr 1974, pp.424–5). A more recent organizational innovation, the 'corporation', represents an attempt to address two of the most critical deficiencies of Soviet industry – the poor integration of R&D activity with the operations of manufacturing enterprises, and the difficulties of arranging flexible intersectoral linkages, which are a necessary element of product development (Gorlov & Baburin 1985). By the end of 1983, corporations accounted for half of total industrial sales and employment. They are organized as hierarchical production systems, with key functions generally concentrated in larger cities and branch plants dispersed among smaller towns. This core–periphery spatial structure is likely to be further entrenched under the policies of *perestroika*: the future of the greater Moscow region is certainly being planned along these lines (Bond 1988). By giving priority to more efficient and concentrated industrial investment and to improved transfer of R&D findings into manufacturing production, Soviet authorities are moving towards a form of growth-pole strategy, focused on the largest metropolitan centres (Bond 1987).

Urbanization, regional development, and social welfare

One of the tenets of socialist ideology is that the geographically uneven pattern of development which characterizes capitalist societies will be eliminated under socialism. At the national scale this will be evident in only minor interregional variations in levels of economic activity and the provision of public services; at the regional scale it will be manifest in markedly reduced urban–rural disparities as compared to those typically found in non-socialist societies; and at the metropolitan scale it will be demonstrated by the absence of the sharply defined socio-economic zonation which has characterized cities in the capitalist world since the Industrial Revolution. Commitment to policies designed to achieve these objectives can be abundantly documented in official proclamations and plans in the USSR. Some of the principles of industrial location listed above are clearly consistent with them. Soviet planners have shown a long-standing interest in the concept of an optimum city size, and in avoiding the social costs of excessive urban growth. It is necessary to ask, nevertheless, how consistently and successfully socialist aspirations have been embodied in the actual distribution of activity, at various geographical scales, which has been achieved during the Soviet era.

The USSR is the world's largest state. In addition, it is one of the few in the world to be faced with three types of 'problem regions':

well populated but industrially underdeveloped regions; older industrialized regions where the economic structure needs rejuvenation and/or
diversification; and harsh pioneer areas where natural resource wealth
prompts attempts to integrate them into the national economy (Schiffer
1985, p.508). This perspective is necessary to a reasoned evaluation of the
levels of interregional disparity which are unambiguously present in the
contemporary USSR. All the available measures of population density,
employment, industrial output, etc. point to the fact that at the national
level European Russia remains the core region of the USSR, and Siberia
and the central Asian republics are less developed peripheries. There is
a growing, if tacit, recognition that this distinction will persist for the
foreseeable future. Especially with respect to Siberian development, the
rhetoric of heroic environmental transformation which typified official
statements in the Stalinist era has given place to a more sober assessment
of the difficulties and costs of investment on the frontier. Moreover, with
a much reduced level of involuntary migration in the post-Stalin years,
the Soviet authorities have found it more difficult to attract permanent
settlers to Siberia, whereas migration to areas bordering the Black Sea
(the Soviet 'Sunbelt') has grown steadily (Shabad 1978). To the extent
that Soviet authorities believe there is an unavoidable trade-off between
efficiency and regional equity in the process of economic growth, they
appear to be placing a greater emphasis on efficiency.

Levels of development and of social welfare are closely correlated
with levels of urbanization in the USSR, as they are in comparable
non-socialist nations. When the Bolsheviks came to power in 1917,
only 18% of the population of the Russian Empire was urbanized.
Rapid urbanization, involving the creation of approximately 400 new
towns, accompanied the industrialization drive of the early Five-Year
Plans, so that by 1939 the Soviet population was one-third urbanized. By
1979 the proportion had reached 62%, which was roughly equivalent to
the position reached in the USA 30 years previously (French & Hamilton
1979). The urban population is regionally concentrated (over 70%) in
European Russia, although the fastest urban growth rates have in recent
years been recorded in the ethnic republics of central Asia (Clem 1980).

There has been noticeable divergence between the principles and
practice of socialist urban development during the Soviet era. Pre-
revolutionary Russia contained two disproportionately large cities, St
Petersburg and Moscow, and the former, in particular, exhibited all the
worst features of uncoordinated capitalist development (Bater 1979).
Early Soviet planners sought to establish an optimum size for new
urban settlements, but over time the theoretical optimum was increased
(from around 50 000 inhabitants in the 1920s to around 250 000 in the
1960s), and the idea appears now to have been abandoned, optima
having been consistently exceeded in practice. The pace of urban

construction (including postwar reconstruction), and the subordination of urban planning concerns to the dictates of industrial ministries single-mindedly bent on expanding their production capacity, have together ensured that socialist aspirations to rational and efficient city development have rarely been fully realized (Bater 1977). To maintain a sense of proportion in identifying these shortcomings, however, one needs only to consider the variety of urban planning problems evident in non-socialist cities.

Disparities between Soviet cities and between metropolitan and rural areas were pithily captured in Premier Khrushchev's statement, 'If it's bad in Moscow, it's worse in the provinces.' An important measure of inter-urban differentials is the amount of living space per capita, which declines steadily from west to east across the USSR. The most pronounced disparities are, however, found at the intra-regional scale. One fundamental cause of low living standards in rural areas 'has been the persistently lower wage paid for work performed in agriculture as opposed to industry and other urban occupations' (Fuchs & Demko 1979, p.308). Despite the measures taken in recent years to redress rural–urban imbalances, rural depopulation accelerated during the 1970s, especially in the long-settled regions of European Russia. Outmigrants have been most attracted to the larger cities which serve as regional administrative centres, and which tend to attract a disproportionate share of new industrial employment (Sagers & Green 1979, Clem 1980). The long-term goal of Soviet planners is to upgrade existing rural settlements into small towns, with a broader economic base and an appropriate range of urban amenities. However, just as the realities of population movement forced constant growth in the specified size of an 'optimum' city, so planners' ideas of the lower limit for an *agrogorod* (agricultural town) had to be adjusted downward (French & Hamilton 1979).

Eastern Europe: variations on a theme

Having stressed that the policies and procedures of the Soviet regime, and the specific environment and challenges which it has faced, are Russian as much as they are socialist, it is worthwhile looking briefly at the socialist states of Eastern Europe. With the exception of Yugoslavia, where the local communists made a smooth transition from being wartime partisans to forming the postwar government, these states began life with a heavy dose of tutelage from Moscow. Especially during the Cold War period (*c*.1948–60), the Sovietization of their economies was pursued strenuously, involving such elements as nationalization of the means of production, collectivization of agriculture, major investments in the industrial sector, and the reorganization of territorial-administrative systems (Mellor 1975). Economic planning was coordinated on a bilateral basis with Soviet plans, and Comecon was created as a vehicle for multilateral coordination. But despite Stalin's efforts to create client states in the Soviet image, it was not long before evidence appeared that socialism in a non-Russian cultural and national milieu could not be guaranteed to reproduce the patterns of the 'socialist fatherland'.

Among the Eastern European socialist states, Poland and Yugoslavia differ most markedly from the Soviet model of agrarian transformation (Fig. 9.4). In both, initial moves to create co-operative (collective) farms were allowed to lapse when the authorities were forced to acknowledge that the process could only proceed on a voluntary basis. By the end of 1956 the proportion of agricultural land farmed by co-operatives had fallen to 1% in Poland (from a peak of 10% three months earlier) and to 2% in Yugoslavia (from a peak of 18% in 1950). State farms have grown slowly to occupy no more than 15% of the agricultural land in each country, but their contribution to marketed agricultural output is considerably greater. As Bergmann (1975, p.172) observes of Poland, 'the continued existence of 3,600,000 small farms and dwarf holdings has posed rather than solved the problem of the country's agrarian structure.' Under different conditions, the Soviet model of reform, implemented intelligently and gradually, was accepted without resistance by the peasant farmers of Czechoslovakia.

Throughout Eastern Europe the socialist regimes have subscribed to the same general principles of industrial location as those followed in the USSR, as reviewed above (Hamilton 1971). Especially in the more developed northern states (the GDR, Poland, Czechoslovakia, and Hungary), major regional concentrations of industry were already established, but the substantial expansion of capacity since 1945 has also involved newly established urban settlements built around industrial complexes such as those of Nowa Huta and Eisenhüttenstadt. In some instances this new investment has helped to lower interregional disparities in industrial development, but the states of Eastern Europe have experienced, admittedly to a lesser degree than the USSR, the tension between production efficiency and regional equity in locating additional capacity (Turnock 1978). Regional decentralization of industry has been pursued most consistently in Yugoslavia, which is unique among socialist countries is having a federal constitution matched by a substantial devolution of decision-making and resource-allocation from the central State administration to the constituent republics. The Yugoslav form of worker self-management which, together with a reliance on regulated market mechanisms rather than directive planning, has given the individual enterprise greater freedom of action than in most communist states, produces some distinctive locational patterns

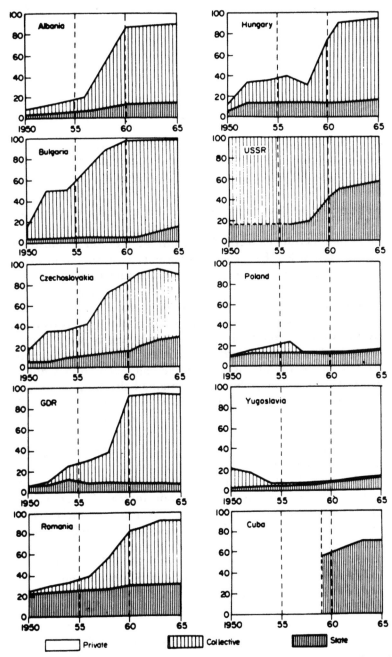

Figure 9.4 Division of agricultural land between the social sectors in selected state-socialist countries, 1950–65.

Source: Bergmann, T. 1975. *Farm policies in socialist countries.* Farnborough, Hants: Saxon House, Figure 1.1

(Hamilton 1974). Currently, it is Hungary that has moved farthest from the rigidly centralized decision-making of a command economy.

The advanced state-socialist economies in the world-economy

The USSR and the socialist states of Eastern Europe together contain one-tenth of the world's population and account for 15% of 'world GNP'. Periods of minimal trading contact with the non-socialist world followed the establishment of the Bolshevik régime in Russia, and the establishment of the European communist states in the late 1940s. Despite the Soviet preference for economic self-sufficiency, the need to acquire more advanced technology from the West once industrialization began in earnest led the USSR to increase its involvement in world trade just as the depression of the early 1930s made such trade very welcome to capitalist enterprises. To pay for imports of machinery, the USSR relied upon grain exports, which were the only saleable commodity the largely agricultural economy could produce in adequate quantity (Karcz 1979).

By the time the political climate of relations between the socialist states and the advanced industrial nations of the West improved in the 1960s, the nature of East-West trade had taken on new dimensions. Advanced production technology remained the most important item on the Soviet, and now also the East European, shopping list, but socialist régimes could no longer be as unresponsive to the welfare and aspirations of their own people as Stalin was in the 1930s. Adverse climatic conditions compounded the production shortcomings of Soviet agriculture in particular, leading to the initiation of imports of grain from the West in the mid-1960s. These have since become a stable element of Soviet bloc imports, to support increased meat production.

The growth in the volume of trade between the industrialized socialist states and the rest of the world has highlighted a number of important questions concerning the future of the global economic system. In the late 1970s, almost 60% of Soviet trade was still with other industrialized socialist nations, but trade with industrialized non-communist states had grown to 30%, and trade with Third World countries stood at about 12%. Overall, the countries of Eastern Europe had a similar trading pattern (the substantial exchange between East and West Germany being an obvious exception). The relatively high level of self-sufficiency of the Soviet economy, and the concentration of its trade among other Comecon members, underlines the limited extent to which the state-socialist economies as a whole are integrated into the capitalist world-economy. However, it is worth exploring the question of whether this reflects a firm ideological stance or a situation forced upon command economies by virtue of their difficulties in performing competitively in capitalist markets.

Growing East–West trade in the 1970s was accompanied by growing Soviet bloc trade deficits. The USSR has been able to limit its indebtedness by increasing its hard currency exports of raw materials, notably petroleum, natural gas, and gold; but the East European countries, lacking the Soviet resource base, have not had this option. By 1981 these states (minus Yugoslavia and Albania) had accumulated a hard currency debt of $64 billion, which forced them to make dramatic cuts in Western imports (*The Economist* 7 April 1984). As events in Poland in the 1980s showed, the political implications of tackling the economic problems underlying this trade position are profound and unpredictable. The USSR also faces major challenges, for further development of marketable resources, notably of energy in Siberia, involves increasingly heavy demands on investment capital and on Western technology. Joint projects with Japanese or European consortia, whereby the foreign partner's contribution is repaid on the basis of a delivery agreement (e.g. for natural gas or coal), were favoured in the late 1970s (Hebden 1980), but the depressed level of world commodity prices in the 1980s has reduced the attraction of such schemes. Indeed, the case for continued investment in resource megaprojects in Siberia and the Soviet Far East (including related transportation infrastructure such as the Baikal–Amur main line) has been largely undermined for the foreseeable future (Bradshaw 1988).

The indebtedness of East European nations and the dominance of natural resources in Soviet exports both point to the relative lack of success of the industrialized state-socialist economies in producing manufactured goods that are globally competitive. Their poor performance is not for lack of skilled labour or basic scientific expertise; rather, it stems from the obstacles to innovation and its diffusion, and the lack of incentives to cost-effective and reliable production in a centralized command economy. Not only are Soviet bloc products, especially consumer goods, generally unappealing to Western buyers; they are increasingly so in Third World markets in comparison to NIC products. The USSR's limited trade relations with the non-socialist Third World have tended to match the pattern of its aid (civil and military), being concentrated on neighbouring states, such as India, and those in the Middle East and North Africa. Among socialist Third World nations, only Cuba stands out in terms of its volume of trade with the USSR – a relationship that is fundamentally shaped by geopolitical considerations rather than economic logic.

Industrialized state-socialism in perspective

The economic geography of the industrialized socialist states is indisputably different from what it would have been had the political and social system prevailing at the time of their transition from a capitalist order remained in place. Abouchar (1979) outlines a comparison, which should not be pushed too far but which is nevertheless suggestive, between Russia and Brazil in 1914. They were at the same level of development per capita and were similar in a number of qualitative dimensions such as the social distribution of wealth and power. Their divergent paths of development since that time cannot be ascribed *simply* to a difference in their political economy, if for no other reason than that their resource endowments differ considerably. Nevertheless, a comparison of their current positions with respect to average income, income distribution, industrial structure, agricultural self-sufficiency, domestic ownership of the economy, and regional disparities indicates something of the 'might have been' had each country followed the other's path to development. It is a measure of what has been achieved in the USSR since the Bolshevik revolution that a comparison with the US, the other 'superpower' (Parker 1972), is not inappropriate.

One can identify a number of contrasting answers to the question of what direction the development of the industrialized socialist economies is taking. Some commentators are most impressed by the convergence of capitalist and socialist responses to the challenges of maintaining a viable modern state. They point to the common characteristics of reliance on technology and bureaucratic organization, and to the similar aspirations towards social wellbeing of populations in each sociopolitical system. On this basis, the solution to similar problems is likely to produce increasingly similar geographical consequences, regardless of ideology. On the other hand, some socialist commentators view the achievements of the USSR and Eastern European states as providing a still imperfect indication of the nature of a truly socialist society. From this perspective, the heritage of capitalist social and economic structures continues to frustrate the realization of socialist principles, not least in areas such as urban planning and agrarian organization: given more time, the radical difference between the geographies of capitalism and socialism will become increasingly evident.

The 1990s are likely to be a significant and fascinating period in the evolution of developed state-socialist societies. Culture and geography both present challenges to what has been orthodox Marxist ideology in a Russian mould. The cultural distinctiveness of Russian society is extremely durable, and it is far from clear how enthusiastically the reforms promoted by Mr Gorbachev will be implemented on the ground in the provinces. The efficiency of a more market-oriented economy is unlikely to be realized without individuals (and individual firms) being giving greater freedom than is currently in prospect, and a greater sense of ownership of their work and means of production. The receptiveness of the non-Russian republics of Soviet central Asia to *perestroika* is even more in question, although some analysts suggest that continuing high rates of population growth will soon be reflected in increased urbanization and 'modernization'. The economies of Eastern Europe, meanwhile, may well become more diverse, both in their internal structure and in their degree of involvement with the capitalist world-economy. Joint ventures between Western TNCs and state enterprises in countries such as Hungary are already serving as one channel of linkage, but the degree to which others develop is likely to hinge on the uncertain evolution of the complex geopolitical relations between the USSR, Eastern Europe, the EEC, and the USA.

10 *The Third World: varieties of underdevelopment*

Theories of underdevelopment

Some basic definitions

No collective term to identify the nations of the global periphery is empirically unambiguous and ideologically neutral. The concept of 'the Third World' has become increasingly unsatisfactory to define that majority of contemporary states which are part neither of the capitalist nor the socialist industrialized core regions. Nevertheless, it remains a widely used and concise designation, adequate in many contexts to convey a clear meaning, so it is used whenever appropriate in this chapter. Any classification of modern states into a limited number of categories inevitably creates problems associated with definitional boundaries. Whether divisions are based solely on quantitative measures, such as GNP per capita, or on qualitative scales of political and economic organization, there will always be some countries in any group which appear anomalous.

While equating the Third World with 'the global periphery', we recognize that core–periphery relationships involve states distributed along a continuum of economic and strategic influence. For this reason, creation of an intermediate category, such as Wallerstein's (1976, 1980) 'semiperiphery', does not fully resolve the classification problem. Peripheral status is always relative to the specific identity of the core with which comparisons are being made. Kenya, for instance, is clearly part of the global periphery, but it functions as a core region with respect to its immediately neighbouring states in East Africa. Moreover, within even as brief a period as the past 20 years, some states' power and prosperity have changed quite significantly.

To refer to the 130 or so constituent political units of the Third World as *less developed countries* (LDCs) immediately poses the question of what constitutes 'development', which is addressed in the next section. Whatever definition one adopts, it is obvious that the level of development varies considerably. In attempting to make some simple but meaningful distinctions, the World Bank has adopted a four-fold grouping of LDCs, into *low-income, middle-income, upper-middle-income,* and *high-income oil-exporting* economies. The growing diversity of Third World states has prompted widespread usage of a number of categories which focus on common characteristics of selected nations. These terms, which cut across the World Bank groups and are not always consistently defined, include the *newly industrializing countries* (NICs), *Third World oil exporters* (of which the OPEC *states are a subset*), and *non-oil-producing Third World nations.* In contrast, one cannot divide the global periphery unambiguously into capitalist and socialist states as one can the global core. Although Cuba represents an economy which is clearly organized on state-socialist lines, there are a number of countries, notably in Africa and Latin America, whose governments espouse socialism, but whose internal social and economic structures and external involvement in the capitalist world-economy exhibit a variety of different forms. Another basis for classifying LDCs, one which is more significant than it may seem at first sight, is to group them by continent.

The nature of development

Goulet (1973) identifies three different conceptions of 'development'. The first holds that, by and large, it is synonymous with aggregate economic growth, as measured by increases in a country's GNP. This narrow definition was dominant during the critical period of the 1950s and 1960s, when today's Third World of independent sovereign states came into being, and when the academic study of development economics began to attract sustained attention. Behind this emphasis on growth lay the recognition that the level of productive output per head in LDCs was very low in comparison to that achieved in the economies of the global core, and strenuous efforts would be required to close the development 'gap'. Rostow's (1960) suggestion that national development in the Third World would proceed through the same stages as that of the UK and the USA was widely accepted (Fig. 10.1); therefore attention focused on the bottlenecks which constrained economic growth. The natural environment and resource base of the LDCs were too readily assumed by conventional economists to present no major obstacles (see p.218); labour was assumed (again, too readily) to be in plentiful supply; so the critical shortage was identified as an inadequate supply of capital to harness the other factors productively. Investment by, and financial aid from, the rich countries of the core would help, but the real challenge of development was to increase the volume of domestic savings: to transform the typical Third World society 'from being a 5 percent to a 12 percent saver – with all the

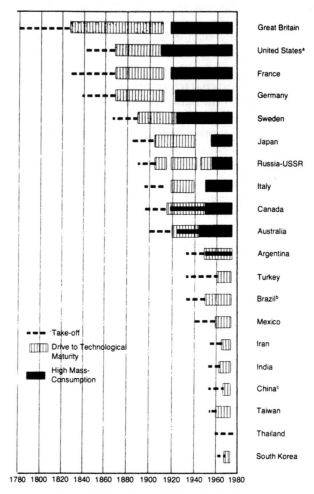

CHART II-2. *Stages of Economic Growth: Twenty Countries*

aNew England regional take-off, 1815–1850.
bSão Paulo regional take-off, 1900–1920.
cManchuria regional take-off, 1930–1941.

Figure 10.1 Rostow's 'stages' of economic growth in selected countries.
Source: Rostow, W. W. 1978. *The world economy: history and prospect.* Austin: University of Texas Press; Chart II-2.

changes in attitudes, in institutions and in techniques which accompany this conversion' (Mehmet 1978, p.19; quoting Lewis 1955).

   Not only did this analysis suggest that the path to development was to maximize national income as quickly as possible, notably by industrialization, but it also implied that a larger pool of savings would be mobilized if the benefits of growth were concentrated within society. An affluent élite has a higher propensity to save than do the poverty-stricken rural masses and, 'in the idiom of the period, there was no point in becoming concerned about [equitable income] distribution when there was practically nothing to distribute' (Mehmet 1978, p.18). Some of the consequences of this conception of development for Third World societies will become evident below.

A second definition, which now commands majority support, sees development as a process involving economic growth, but also requiring qualitative changes in a society's social and economic life. This perspective gained favour during the 1960s and 1970s, in part because of a growing awareness that the narrower definition just described not only failed to acknowledge the multidimensional character of Third World aspirations, but was resulting in strategies of economic development which were not proving successful even in their own growth-oriented terms. A common feature of conceptions of development which define it as 'more than growth' is that they embody a concern for equity. One expression of this focus is the *'basic needs' approach*, which argues that a major obstacle to sustained economic growth in most LDCs is the fact that so many people lack adequate food, shelter, clothing, and access to even rudimentary provision of uncontaminated water and preventative health care. If public investment to meet these needs were given priority, there could be a radical and far-reaching improvement in the economic productivity of the population (see p.271). A related concept is that of *'bottom-up development'*, as opposed to a 'top-down' approach (Stöhr & Taylor 1981). Whereas policies designed to maximize aggregate economic growth tend to promote centrally planned, capital-intensive investment, the benefits of which are generally confined both socially and geographically to an urban-based élite and tend not to 'filter down' to the rural masses, bottom-up strategies give priority to developing the human and resource potential of the entire nation through the creation of regional economies which are relatively self-sufficient and in which the planning process is locally directed (i.e. decentralized) to the greatest feasible extent.

Goulet's (1973, p.xiv) third conception of development, whose advocates have 'always constituted, in some respects, a heretical minority', explicitly focuses on its ethical dimension. The value-orientation of the basic needs approach already indicates the degree to which a definition of development as simply economic growth or 'modernization' has yielded place to a fuller appreciation of development as a human enterprise. Nevertheless, debate about development still tends to focus more on the ends than the means, or on the process rather than the individuals whose lives are changed. Berger (1974, p.xi) argues that 'underlying the major ideological models for social change (including Third World development) are two powerful myths – the [capitalist] myth of growth and the [socialist] myth of revolution. *Both* myths must be debunked.' Having reviewed the very different development strategies of Brazil and China, he finds they exhibit a common ethical bias, a willingness to sacrifice the life of the individual for the sake of a theory of society. Hence he calls for solutions to the problems of development which 'accept *neither* hunger *nor* terror', and which allow all those who are directly involved to participate in defining what those 'problems' are. Both dynamic economic growth *and* economic stagnation entail significant human costs. Once the myths of development are dispelled, individuals can face existential choices of policy and action more perceptively, aware that the outcome of any major social change is necessarily uncertain in advance.

The fundamental theoretical question posed by the underdevelopment of the global periphery is whether LDCs should be viewed as prospective core states, or whether their characteristics are a direct consequence of the 'development' achieved in the core. If the Third World is potentially capable of enjoying the levels of per capita income achieved in the industrialized nations, and if this prosperity depends upon social, political, and economic changes which are essentially universal in their applicability, acceptance, and outcome, then, 'in the end, late-comers to modern growth can, indeed, catch up with early-comers . . . [and] it is not an impossible dream to envisage their coming to a recognizable version of mass affluence at different times over the next century or so' (Rostow 1978, p.656). If, on the other hand, 'it is in the underdeveloped world that the central, overriding fact of our epoch becomes manifest to the naked eye: the capitalist system, once a mighty engine of economic development, has turned into a no less formidable hurdle to economic advancement' (Baran 1957, p.402), then prospects for improving the wellbeing of the majority of the world's peoples are poor (in the absence, Baran would add, of socialist revolutions) and change will be slow.

These opposing diagnoses of the condition and the future of the Third World are, in their ideologically embellished forms, precisely those myths of growth and revolution which Berger argues must be debunked. The evident diversity of contemporary LDCs is, in itself, an indication of the complexity of the development process. Theoretical frameworks, whether derived from neoclassical or Marxist economics, which reduce this reality to the operation of a few variables may 'satisfy both professional elegance and emotional needs' (Seers 1981, p.146), but their apparent clarity and comprehensiveness masks their weaknesses when confronted with specific contexts. The significance of many geographical and cultural variables is notably discounted. In particular, the external orientation of the State, its domestic policy priorities, and its administrative competence and integrity (see Ch.5) fundamentally shape its development achievements and prospects.

Dependency theories
The position of peripheral states within the capitalist world-economy has attracted a number of *dependency theory* interpretations. The characteristics and dynamics of dependent development are seen to arise because peripheral countries are 'conditioned by' (Dos Santos 1973, p.76) the nature of economic activity initiated in and propagated from the core. Differences of opinion among dependency theorists hinge, very largely, on varying interpretations of the conditioning process (Palma 1981, Corbridge 1986). There is considerable disagreement, in the first place, as to the degree of determinism it involves. Whereas some writers, such as Frank, describe an almost automatic process whereby the core moulds the periphery into an antithesis of itself ('the development of underdevelopment'), others recognize a powerful, but contextually varied, influence which the existence of the core economies has on the development prospects of the periphery. Secondly, differing assessments of determinism reflect differing evaluations of the role of *internal*, as opposed to *external*, factors in the creation and maintenance of dependency relationships. The more mechanistic theories argue that influential groups in the periphery will *inevitably* identify their interests with those of the capitalist enterprises which integrate their domestic economies into the core-dominated world-economy. Other theories, however, take the development problems posed by the extensive penetration of peripheral economies by core agents (such as TNCs) as the starting point for defining strategies whereby national governments can free themselves of the less desirable impacts of core domination.

Dependency theories have been significantly shaped by two specific sets of national development challenges. In the early years of the 20th century, the problems facing Russia as a latecomer to capitalist development attracted the analytical scrutiny of authors such as Lenin and Rosa Luxemburg. Among the salient characteristics of pre-revolutionary Russia was that its social structure differed significantly from that of the core states of Western Europe, both in the relatively small size and weak position of its entrepreneurial middle class, and in the relatively large size of its traditional, 'pre-capitalist', peasantry. As a result, industrialization and related forms of economic development relied heavily on foreign capital and technology, and Russia's attempts to develop along the lines pursued by the UK, France, Germany, and the USA were hindered because its fledgling modern manufacturers had to compete not only with the output of traditional industries at home (as had early manufacturers elsewhere), but also with that of far more efficient, and increasingly powerful, producers exporting worldwide from their factories in these core states. The Soviet response to these challenges was reviewed in the previous chapter, and some contemporary dependency theorists imply that only a Soviet-style socialist revolution and programme of state-directed economic development will meet the current needs of Third World countries which exhibit many of the characteristics just described.

A group of writers associated with the United Nations Economic Commission for Latin America (ECLA) analyzed the development challenges facing the nations of that continent (see Brookfield 1975), and in so doing elaborated upon many of the issues raised by Lenin and Luxemburg. Starting from a critique of the Hecksher–Ohlin theory of international trade (see Ch. 3), a basic claim of the ECLA analysis was that the structure of production in the periphery differs substantially from that in the core. Output in the core is 'homogeneous and diversified', whereas in the periphery it is 'heterogeneous and specialised' (Palma 1981, p.51). What these terms point to is that core economies are characterized by a relatively uniform level of productivity in the various branches of production and by intricate intersectoral linkages which sustain a wide variety of final outputs. In contrast, peripheral economies are marked by substantial differences in the level of productivity (notably between the 'modern', primarily export-oriented, sector and the 'traditional' sector) and by the concentration of marketed output in a narrow range of industries, the operations of which have very few linkages with each other or with the rest of the domestic economy.

This structural difference between the core and the periphery has a very powerful influence on the dynamics of the interaction between them within the capitalist world-economy. Over time, the pattern and pace of economic development experienced in the two regions diverges, for the following reasons:

(a) The periphery's dominant traditional sector is hard pressed to generate technical progress or to integrate it into productive activity. As a result, labour productivity (and earnings) rise more slowly in the periphery than in the core.

(b) The periphery's low-productivity sectors, especially subsistence agriculture, continuously generate an excess labour supply which exerts a strong downward pressure on wage levels in the modern sector. This adversely affects both the level of domestic demand for goods production and also the prices of exported goods, so that it becomes a factor responsible for deterioration in the periphery's terms of trade with the core.

(c) Core–periphery differences in productivity growth and in the terms of trade lead to a growing divergence in real income levels, and so to an increasingly polarized pattern of world development.

Can peripheral nations break out of this apparently dismal trajectory? The ECLA analysis suggested that this can be achieved if governments take steps to transform their domestic economies. One vital initiative is to foster the development of *import-substituting industries* so that, as the economy grows, an increasing proportion of manufactured items is produced at home rather than imported. To achieve this, the state will need to intervene to protect 'infant' industries from foreign competition (on the lines of the tariff policies initially adopted by the USA and Canada to establish their domestic manufacturers), and will need to invest directly in critical, capital-intensive sectors, such as iron and steel, and electric power generation. At the same time, it must promote as vigorously as possible a broad and sustained growth in domestic real income, which implies a conscious effort to raise productivity in the traditional rural economy.

As we note below, this specification of the transformations required to promote Third World development is well conceived, and the achievements of the NICs indicate that it can be successful *under certain conditions* (which practically and theoretically remove it from a 'dependency' perspective [Browett 1985]). However, as Palma (1981, p.55) notes, the weakness of the ECLA analysis lies in the fact that 'no consideration is given to the social relations of production which are at the base of the process of import-substituting industrialisation, and of the transformation in other structures of society that this brings in its wake.' In particular, Lenin's focus on the ambiguous rôle which foreign investment, and the domestic social groups whose interests became associated with it, play in the development process is pertinent to any analysis of the Latin American experience. It has also been applied to interpret the experience of the more recently independent African and Asian countries (Weinstein 1976). In terms of Figure 5.2, where attitudes associated with the lower right quadrant are prevalent among those in power in Third World states, dependency relationships are more likely to be entrenched than to be challenged.

Development is a process, and however individuals or governments define the goal, their theory and practice are intimately interwoven. Therefore, we turn now from the discussion of conceptual issues to a review of the objective conditions which a developing nation must address.

The context of underdevelopment

Introduction
The evolution of development studies among European and North American academics has been shown to exhibit ethnocentricity and cultural paternalism with respect to Third World societies (e.g. Berger 1974, Nafziger 1976, Goulet 1980). Writers from quite different philosophical backgrounds have given the impression that

societies of non-European origin have been the passive recipients, or victims, of 'development' initiated by representatives of the core states. The behaviour of tropical agriculturalists has been interpreted unsympathetically by authors who have implicitly subscribed to the 'Protestant work ethic' and have been complacently ignorant of the environmental challenges facing cultivators in non-temperate latitudes. Even amongst interpretations of underdevelopment offered by Third World writers, there has been a tendency to use categories of analysis derived from the dominant core ideologies. The following analysis of contemporary Third World development will attempt to address the rôle of the natural environment, the influence of past events, and the perceived opportunities of the present. It is impossible to estimate 'objectively' the relative significance of each factor in explaining underdevelopment, but it will emerge that theoretical frameworks which focus exclusively on one are seriously misleading.

Environment and resources

If a past generation of geographers was guilty of environmental determinism (see Ch. 2), too many economists have been guilty of environmental naïvety. Neoclassical theory which takes 'land' as a given, and Marxian theory which attributes all economic output ultimately to human effort, encourage the belief that variations in environmental conditions in different parts of the world have little or no influence on the course of economic development. Only gradually has it become widely appreciated by development specialists that theories developed in the temperate core regions of Europe and North America cannot be assumed to be valid in the very different climatic and soil conditions of the tropical world. Grigg, a geographer, very appropriately entitled his study (1970) of the problems facing Third World farmers *The harsh lands*. More recently, a World Bank economist felt it necessary to remedy his profession's past neglect of the rôle of climate in economic development, because its effects impinge directly on productive activity and beyond into such areas as nutrition, health, and education (Kamarck 1976).

Europeans' initial assumption that the luxurious vegetation of the humid Tropics was evidence of a highly productive natural environment was only half correct. Certainly, the amount of solar radiation received in these regions is between 60 and 90% more than that received in the humid temperate zone, and organic productivity is correspondingly greater. However, the dynamics of the ecosystem which supports this growth were long misunderstood. Nutrients are recycled at the surface of tropical soils, but below the surface layer their organic content is generally poor. Heavy rainfall promotes sustained *leaching*, whereby plant nutrients are removed from the upper soil horizons; and high temperatures hasten the formation of *laterite*, a hard, infertile soil with

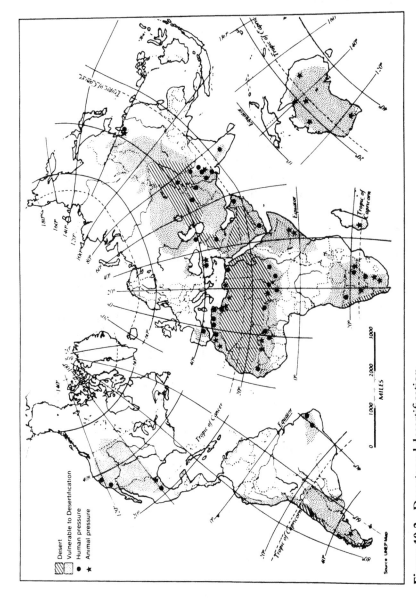

Figure 10.2 Deserts and desertification.
Source: U N Environment Program.

heavy concentrations of iron and aluminium oxides. The most productive soils in the humid Tropics are therefore derived from geologically young materials, on which these processes have had limited time to act. This means that alluvial soils and those derived from recent volcanic deposits are particularly significant, and Southeast Asia stands out as being well endowed with both (Grigg 1970).

Many tropical regions suffer from notably unpredictable rainfall, and the threat of desertification on their arid margins has attracted growing attention since 1970 (Fig. 10.2). Desertification is both a human and a natural phenomenon, for it involves the precarious and complex interaction between populations and a resource base, the productivity of which is subject to pronounced year-to-year variation. Population growth, together with the penetration of a market economy and of new technologies (such as artesian wells) in ways which disrupt traditional practices and result in 'the loss of livelihoods and local wisdom with proven viability' (Hare et al. 1977, p.341) are part of the human dimension.

Those who live in regions vulnerable to desertification tend to be 'marginal' in every sense of the word – poor, powerless, and remote. Ferraro et al. (1982, p.7) argues that 'poverty restricts choices concerning resource use', and this helps to explain why the degradation of semi-arid environments continues, even though the populations involved are directly aware of the rising costs which their practices impose on themselves. Nowhere is this more evident than in the continuing deforestation which results from a dependence on fuelwood as the only affordable energy source. Whether measured in terms of monetary or of labour (collection) costs, the price of fuelwood is rising rapidly, and there is a growing deficit of supply, which in 1980 affected one billion users, predominantly in rural areas. Loss of the ground cover and root systems which trees provide hastens the destruction and removal of the soil, making regeneration of the productive capacity of the land extremely expensive, if not impossible.

The health of human and animal populations in most of the Third World is strongly related to environmental conditions in ways which re-emphasize the difference between the temperate and tropical worlds. In the absence of a winter season which suppresses weeds and pests, tropical populations and cultivated plants are subject to the impact of an infinite variety of continuously growing life-forms which threaten their wellbeing or productivity. Pre- and post-harvest crop losses can amount to 50% of potential production; yet the cost of counteracting disease and predators can amount to almost the same proportion of the value of crops (Kamarck 1976). Staple subsistence crops are generally less vulnerable than exportable cash crops, but livestock is uniformly at risk. Africa suffers to a much greater extent than Asia and Latin America from diseases which kill or debilitate cattle, horses, and other draught animals, which helps to explain the slow and late penetration of the continent by Europeans. Parasitic diseases are widespread in tropical regions, especially in Africa, inflicting debilitating illnesses on people whose poverty and generally poor levels of health and nutrition are related to them by a chain of circular and cumulative vulnerability. Combatting these diseases requires that financing, appropriate infrastructure, and international coordination be available *simultaneously* for a prolonged period. Inadequate resources and the social and economic dislocations of armed conflict and civil strife help to explain why so little progress (indeed, retrogression) is evident in their eradication from parts of the contemporary Third World.

The Third World's endowment of non-renewable resources presents a more complex geographical pattern than that associated with generalized environmental conditions. A number of points need to be made in order to assess the development implications of these resources correctly (Chisholm 1982). First, despite the increasing sophistication of remote sensing and aerial surveys, large parts of the Third World have yet to receive a basic mapping and inventory of their resource base. Undoubtedly, LDCs are better endowed than currently appears. Secondly, however, it is clear that, with a few exceptions, the possession by Third World nations of exportable surpluses of fuels and minerals is no guarantee of prosperity. We have already noted the unique place of oil and of the OPEC nations in the global resource economy (see Ch. 7). The market for metal ores is very different, partly because reserves of the most important industrial raw materials are widely distributed, and plentiful in core regions (taken as a whole). Thirdly, apart from China and India, the Third World is remarkably lacking in coal deposits. However, Africa, with 27%, Latin America, with 20%, and non-communist Asia, with 18%, possess two-thirds of the world's hydroelectric power potential (Cook 1976).

The size and location of states
Seers (1981) argues that the nationalistic and Marxian perspectives of dependency theorists both tempt them to ignore some basic geographical and demographic differences between states which significantly affect their policy options. Size (considered here primarily as territorial extent – population is reviewed more fully below) is obviously of major importance, if only because it is associated with the quantity and variety of natural resource endowments available to be harnessed. States such as Brazil and Malawi are bound to develop economies which differ in their internal diversity and dependence on foreign trade. Industrialization in small states is often constrained by technological requirements which mean that efficient production is associated with a minimum plant size

well in excess of domestic needs. The global location of states, involving specific geopolitical and core–periphery relationships, also affects their development prospects. An African regime pursuing policies of 'popular socialism' is likely to face much less external interference than a Central American one. Strategic considerations appear to have governed the distribution of US food aid to Third World countries more than evidence of 'objective need' (Tarrant 1980). Regional location, particularly if a nation is landlocked, may severely constrain development options, and boundary disputes may impose a constant drain on a government's attention and resources.

The Population/resource ratio

Since about 1920, when Third World nations began to undergo the demographic transition, exponential population growth has become perhaps the most significant single influence on their development prospects. In comparison to the core regions of Western Europe and North America, whose dramatic population increase took place in the 19th century, the Third World suffers on two counts (Bairoch 1975). Its rate of population growth, especially during the 1950s and 1960s, has been very much higher (peaking at about 2.4% per year in 1965, as against a peak of 0.7%); and the level of agricultural productivity supporting this growth has been very much lower. (Bairoch estimates that in the 1960s productivity in the Third World, excluding China, averaged only 40% of that achieved in British agriculture in 1810.) Whether we view the pressure of incessant population growth on the resource base primarily in terms of its implications for food supply, or for the ability of the agricultural sector to generate a surplus and thus promote economic diversification, it is clear that contemporary LDCs face a much more difficult challenge to raise the living standards of their populations than did the core nations at a comparable stage of development.

The dynamics of continuous population growth in agrarian societies which are, for all practical purposes, confined to a specific and limited resource base, suggest that in the absence of agricultural intensification (which does not offer infinite relief), one may expect to see the falling average output associated with agricultural involution (see Fig. 2.3). In the context of the contemporary Third World, this prompts two observations. First, there is considerable variation in the population : resource ratio between continents as well as between individual states, and this is an important factor in explaining the relative prosperity of Latin America. Secondly, one cannot ignore the influence of the world's sociopolitical order, past and present, on its natural endowments. Harvey (1974) demonstrates the degree to which many neo-Malthusian interpretations of the poverty of the Third World, which stress the dire but 'natural' consequences of overpopulation, serve ideological interests in deflecting attention from aggravating factors such as the grossly unequal access to agricultural land among many Third World populations. These issues are dealt with at greater length in Ch. 11.

The colonial inheritance
Nobody knows how Third World societies would have evolved if they had never been colonized by imperial European powers. Moreover, there was such a variety in the level and character of pre-colonial economic development, and in the duration and impact of colonial rule in different parts of the Third World, that it is extremely difficult to assess the *difference* which colonialism has made to the economic development prospects of contemporary LDCs. Hence Hopkins (1978, p.5) criticizes both dependency theories, which stress the *damage* caused by the penetration of capitalist imperialism, and modernization theories, which accentuate the *benefits*, as being 'heavy with ideology, thin on evidence, and heroic in their assumptions.' Like every other major movement in human history, colonialism reflected the fundamental amgibuities of human nature: a mixture of altriusm and crude lust for power or profits; of well meant, but insensitive, cultural displacement and blatant racism; of improvements in the welfare of native peoples and their unchecked impoverishment. We may decide that, on balance, colonial rule and the economic policies which were imposed along with it have widened the gap between the global core and the global periphery, but we need to recognize that ascribing the relative poverty of the Third World to these factors alone is to obscure the real complexity of the development challenge facing its peoples.

The importance of examining specific contexts of colonial impact can be clearly demonstrated by a comparison of British West Africa and Latin America. The formal colonization of parts of West Africa (modern Ghana and Nigeria) by Britain lasted less than 100 years (1870–1960) and was imposed on indigenous societies which had already developed a market-oriented trading economy. Cash-crop production was predominantly in the hands of native smallholders, and native manufacturing and financial institutions emerged and managed to survive despite the preferential treatment of British businesses (Hopkins 1973). Milewski (1981, p.115) notes that, from as early as the 1890s, nationalist leaders in Nigeria were primarily entrepreneurs, 'struggling not for economic freedom in the abstract but for the freedom to develop Nigerian, national capitalism'. In contrast, most of Latin America experienced 300 years (1520–1830) of Spanish and Portuguese colonialism which created authoritarian, feudal societies in which the natives were firmly subjugated and their economic initiatives suppressed. So strong was the impress of the complementary authority-structures and world-views of the (Roman Catholic) Church and the colonial administration that they have conditioned patterns of development in the century and a half since formal colonization ended. Even in the 1970s, it could be said of manufacturing firms in São Paulo, Brazil that:

> except for foreign concerns, business is run as a feudal extension of the extended family. The *patron* surrounds himself with 'men of confidence' (relatives) and retains company stock within the family. He pays extremely low wages, attempts to take inordinately high profits, and reinvests at a fairly low rate. The important goal of the São Paulo industrialist is not the success of the business *per se* but the amount of status that accrues to him as the *patron*. He views the national government as a private service to protect him from competition (Littwin 1974, p.4).

The Latin American colonial experience is the primary context for Frank's (1979, pp.153–4) formulation of 'the apparently inexorable law of world capitalist development that riches, through the exploitation they invite, generate the development of underdevelopment' (see p.71). In other parts of the Third World, however, the damage done to indigenous economies by colonial rule has been identified more with the distorted pattern of development which it promoted than the systematic extraction of natural resources which it encouraged. McCarthy (1982), for instance, suggests that in Tanganyika (modern Tanzania), the expansion and self-maintenance of the colonial bureaucracy was as great a drain on the indigenous economy as the mechanisms of external trade. The specific character of underdevelopment fostered by the 'bureaucratic economy' involved suppressing grassroots horizontal economic integration, stifling the growth of domestic credit and money markets, misallocating factors of production (including native entrepreneurship) by edicts which made life easier for the administration at the expense of the actual needs of producers, and distorting the communication of economic information by relying sometimes on market signals, sometimes on official decree, and often on both. In the light of such evidence, it has been suggested that the underdevelopment created by capitalist imperialism was as much a product of the comprehensive fettering of indigenous responses to market incentives as of the unfettered penetration of market forces into the global periphery.

Development prospects

Core–periphery relations
The development options facing the contemporary Third World are constrained by external forces, notably the passage of 'world time' (see Ch. 5) and the structural evolution of the capitalist world-economy.

Nevertheless, governments and peoples retain substantial freedom of choice in deciding the nature and priorities of their domestic development efforts. Neither the successes nor the failures of the post-Independence years can be adequately explained simply by reference to the legacy of the past or to quasi-mechanical external influences, such as the 'benefits' of a 'liberal world trading regime' or the 'pernicious consequences' of 'capitalist penetration'. In the periphery, as in the core, individuals and institutional decision-makers give shape to their society by making a myriad of choices for which they are accountable, in the situations which face them. A detailed review of domestic development challenges follows in the next chapter.

But what are the *external* dimensions of the situation which faces the Third World? One of the most noticeable features is the marked disparity of wealth and power between the nations of the global core and those of the periphery. Within Western core states, the threat to national cohesion posed by pronounced interpersonal and interregional income disparities has been alleviated by various forms of wealth redistribution, some of them embodied in explicitly spatial policies. The socialist core states also implement measures to mitigate geographical variations in wellbeing. At the global scale, however, there is no equivalent of an effective national government responsible for promoting a reasonably equitable distribution of society's wealth. It therefore becomes a critical question whether, in the absence of politically sanctioned mechanisms of redistribution, the various forms of core–periphery interaction increase or reduce the differential in levels of wealth and power between the industrial nations and the Third World. Flows of trade, investment, unconditional grants, and the migrations of skilled personnel are among the most significant elements of this interaction. Is their net effect to promote or to hinder the development prospects of the LDCs?

Trade
One of the objections to neoclassical theories of international trade based on the concept of comparative advantage is that they address neither the long-term desirability of a particular specialization nor the implications of the institutional framework within which trade takes place (see Ch. 3). Indeed, differences in the structure of output and in prevailing productivity levels between core and peripheral economies are likely to widen the income gap between them. Underlying this assumption is the phenemenon known as *Engel's Law*, which points to the fact that as incomes rise, the proportion of expenditures devoted to basic necessities (primarily food) tends to decrease. In contrast, purchases of the majority of manufactured goods and of specialized services tend to increase disproportionately, reflecting their high *income elasticity of demand*. Peripheral states which specialize in the export of primary,

especially agricultural, products will thus tend to experience a much less dynamic growth in the *quantity* of demand for their output than will the industrialized core nations. Moreover, for reasons noted earlier, the *price* of primary exports from the periphery is likely to fall over time relative to the price of manufactured exports from the core (see Ch. 3). The *commodity terms of trade* provide a ratio measure of the price of a nation's typical unit of exports compared to the price of a typical unit of imports. From what we have just argued, one would expect the periphery's terms of trade with the core to deteriorate gradually, implying the need to *increase* its volume of exports to finance a *constant* volume of imports; but the commodity composition of peripheral exports suggests that there is limited scope for such adjustment. The logic of promoting import-substituting manufacturing in the periphery becomes apparent in this context.

Considerable disagreement (fed by divergent ideological interests) exists over whether the Third World, as a whole, *has* suffered from a deterioration in its terms of trade. Alternative sources of data, variations in their geographical or commodity coverage, and variations in timespan of analysis are all potential sources of incompatibility between different studies. There is unanimous recognition that, during the quarter-century of rapid growth in the global economy up to 1972, the terms of trade moved against primary producers. They then improved, partly as a result of OPEC actions, but deteriorated once more in the 1980s (see Ch. 3). Controversy arises when one adds a historical dimension, for Bairoch's careful analysis (1975) of data for the period 1870 to 1938 suggests strongly that there was a steady and substantial improvement in the terms of trade between primary products and manufactured goods (which may have begun as early as 1796). Was the experience of the 1950s and 1960s, to some extent at least, a 'correction' of the unusually strong position of the primary producers in 1950? Obviously, there is no objective basis for establishing (nor for refuting) such an argument, but simply to pose it achieves two useful purposes, as Bairoch notes. On the one hand, it suggests that sweeping claims about the inevitability of 'unequal exchange' in the world trading system leading to the impoverishment of the periphery need to be very carefully scrutinized. On the other hand, it indicates that even on the most conservative of assumptions, the loss of purchasing power which the Third World experienced during the 1950s and 1960s negated much of the value of the various forms of development aid which it received from core nations during that period. In the 1980s, its debt repayment potential was steadily undermined by worsening terms of trade.

Loehr & Powelson (1983) summarize a number of studies which survey terms of trade data up to the later 1970s, and reinforce the conclusion that there is no evidence of *long-term* deterioration in the relative level of primary product prices compared to those of manufactured goods. However, there is undoubtedly greater instability in the price level of primary products, which tend to enjoy short periods of rapidly improving terms of trade, followed by long periods in which the price of manufactures exhibits relative gains (Colman & Nixson 1978). The instability of export earnings imposes greater costs on LDCs than would a more stable price régime, and it provides a rationale for attempts to negotiate international commodity agreements which would reduce it. Quite clearly, specific commodities (and hence the specific countries which export them) *have* suffered a fairly consistent deterioration in their terms of trade. Non-cereal agricultural commodities, such as bananas, tea, and sisal, have been especially vulnerable. Oil, and oil-exporting nations did, of course, enjoy a dramatic improvement in their terms of trade for the decade after 1973. This has now reversed, although it is expected to reappear early in the 21st century. Other primary products, such as metals, have also experienced a chequered history in recent decades. It becomes apparent that the heterogeneity of national development achieved in the contemporary Third World has been, and will continue to be, partly a reflection of the market prospects for specific primary products and the LDCs' degree of success in diversifying away from a *staple trap* position (see Ch. 3).

The performance of the NICs since the early 1970s points to the limitations of theories of circular and cumulative change in periods of major structural shifts, and of core–periphery trade models based on a simple exchange of manufactured goods for primary products. New technological and managerial capacities which have made possible a 'new international division of labour' have given some Third World states the chance to capitalize on their newly exploitable comparative advantage for labour-intensive manufacturing. Their attempts to do so, and the freedom of other LDCs to emulate them have, however, been curtailed by restrictive trading measures imposed by the core states of Europe and North America. (Note that the quantity of manufactured goods which the socialist core states purchase from Third World producers is minute). Textiles and clothing were the first industrial sectors in which the divergent interests of core and peripheral nations became obvious, and in which patterns of dealing with 'disruptive' trade (as core governments regard it) became established (Steed 1981, Woolcock 1982).

The labour-intensity and relative ease of entry of these sectors make them obvious starting points for Third World industrialization, as they were for Britain's in the 18th century. At the same time, they continue to employ a significant proportion of the labour force in Western core states (12% in 1975), often concentrated in regions offering limited employment opportunities. Encouraged by political pressure from management, unions, and regional interests, core governments have attempted to preserve domestic jobs by limiting clothing imports from low-wage producers. Woolcock's observation (1982, p.46), that 'neither a majority of [core region] governments nor industries have been prepared to accept that even regulated trade in textiles and clothing would eventually result in a decline in the size of the [core region] industries', can be increasingly applied to other industrial sectors, ranging from steel and ships to colour television sets and automobiles. The unwillingness of core societies to face the adjustment costs of accommodating the potential growth of industrial exports from LDCs is perhaps the most serious external impediment to contemporary Third World development. Admittedly, 'such costs are by no means trivial, even in macro-economic terms, let alone to the individuals most adversely affected' (Steed 1981, p.290); yet one needs to balance that assessment with a recognition of the cost of restrictive trade to individuals in the periphery, whose wellbeing is even less protected by the State, and of the potential benefits to industrialized core nations of their increased exports to more prosperous LDC markets.

Investment, aid and migration

In addition to their export earnings, LDCs have a number of supplementary sources of income to finance their development programs. Each one has characteristics which affect its impact on the domestic economy. Financial resource flows totalling 80 billion dollars entered the Third World in 1978, nearly all of which originated in the global core. (Small but relatively generous aid flows from high-income members of OPEC were the principal exception.) Thirty per cent consisted of *official development assistance*, which includes outright grants from governments or international agencies, or else concessional loans; a very small proportion represents private grants from voluntary organizations. The remaining 70% consisted of various commercial (non-concessional) transactions, the majority of which involved bank loans, foreign direct investment by TNCs, and non-governmental export credits (UN 1983a). From the point of view of the recipient, grants are obviously the most desirable form of development assistance. The value of concessional loans depends on the institutional channels of their delivery. For instance, *tied loans* require that the LDC authorities spend most or all of their receipts in the donor country. This may involve the purchase of equipment which is not the most suitable, technologically, for the domestic economy, and which may cost up to 20% more than if the recipient had been free to 'shop around' for equivalent supplies from elsewhere (Colman & Nixson 1978). Despite periodic commitments by core-region governments to increase the proportion of their GNP allocated to official development assistance, their record has (with honourable exceptions such as the Scandinavian countries) been disappointing (Brandt 1980). Between 1970 and 1978 the proportion of financial resource flows to the Third World composed of grants and concessional loans fell from 45 to 30%; meanwhile, commercial bank lending almost doubled its share, from 15 to 28% (UN 1983a).

Loans incur interest payments, and foreign direct investment generates a stream of repatriated profits, to the extent that Griffin (1970b, p.100) has argued that 'capital imports, rather than accelerating development, have in cases retarded it'. Obviously, the value to a Third World state of infusions of capital from outside will depend largely on the uses to which it is put. The slowing of economic growth in core states in the 1970s, and hence the reduced export earnings of most LDCs, together with the rise in world oil prices (which benefitted only a few of them and raised import costs for the rest), greatly increased the difficulty which many Third World countries experienced in attaining a positive *trade balance* (i.e. export earnings minus the cost of imports) and generating investment capital. In the long run, borrowing is an option only for states whose economic growth is reasonably assured. Some of the very poorest LDCs have proved to have so little hope of ever paying off their debts that a number of lender nations have cancelled them. The future prospects of these impoverished states, many of them in Africa, are not bright, for their development requires a more effective administration than recent regimes have been able to provide and a greater volume of concessional aid, on generous terms, than core nations have been willing to give.

Relatively speaking, foreign direct investment and commerical bank loans have been concentrated in the more prosperous 'middle-income' LDCs. These nations, rather than the very poorest, have accounted for the Third World 'debt crisis' of the 1980s (George 1988). National indebtedness in the early stages of economic development is not, in itself, a serious problem. As long as the borrower can finance the service charges on the debt from export earnings and use the capital to build up its productive capacity, it can accelerate its growth considerably. This is precisely how colonies of settlement, such as the USA and Canada, financed their expansion in the 19th century. However, if loans are mis-spent (which is an endemic problem in 'soft states') or sunk into ineffi-cient para-statal enterprises; or if the investment by TNCs supports non-competitive, import-intensive production for the domestic market (with the steady repatriation of monopoly profits); or if resource exploitation is accompanied by the inequitable sharing of revenues which the major oil companies imposed on host countries for a long time (see Ch. 7), then Griffin's argument about the impact of capital imports carries weight. But even Third World governments which act wisely to minimize most of the potentially harmful side-effects of foreign indebtedness on their domestic economy are at risk – their margin of error or of defence against unpredictable events is very thin. Many of their difficulties in the 1980s have stemmed from the rise of interest rates on global capital markets, for which the growing budget deficit of the US government has been largely to blame.

There remains the basic external constraint that when core-region lenders, especially the commercial banks, have been granting loans to rapidly expanding LDCs (notably the NICs) to purchase capital equipment from TNCs, core–region governments have been simultaneously imposing severe import restrictions on the manufactured goods with which the borrowers could earn the funds to repay those loans. (The problems of countries such as Mexico and Nigeria were compounded by import-intensive investment programmes planned on the basis of over-optimistic forecasts of oil revenues.) This situation, which manifests one of the 'contradictions' of contemporary capitalism, arises partly out of the conflict of interest between TNCs and their home governments (Helleiner 1982). It illustrates the degree to which core states have the power to solve (or at least to attempt to) their domestic economic problems at the expense of weaker states in the global periphery. The serious implications of this relationship are reviewed in Chapter 12. Insofar as international lending agencies influence the domestic policies of Third World debtor states, they frequently reinforce the tendency of ruling élites to act in ways which are detrimental to the interests of the poor (Hayter 1971, George 1988).

Capital is not the only factor of production which is internationally mobile, and movements of people are also a channel of core–periphery relations. Whereas the transition from an agrarian to an urban industrial society was made easier in 19th–century Europe by the available option of mass emigration to 'empty' colonies of settlement, no comparable opportunities are open to the populations of the contemporary Third World. Nevertheless, some states have benefitted from the temporary migration of nationals who work abroad in higher-income countries and remit a portion of their earnings home. Western Europe has long been the principal source of employment of this nature, but the oil-exporting nations of the Middle East also became significant during the 1970s. In 1978, a number of states received remittances equal in value to over half their merchandise export earnings (e.g. Egypt 89%, Pakistan 93%, and Burkina Faso 60%). However, this source of income is unstable and only benefits states which are adjacent to the major markets for migrant labour (World Bank 1981).

A somewhat different aspect of core–periphery labour movements is the international 'brain drain' (Seers 1970). Although the provision of higher education, often on concessional terms, to Third World nationals by core–region institutions can make a positive contribution to upgrading the 'human capital' of developing societies, it can also have quite negative consequences. During the 1950s and 1960s especially, when immigration to North America and Western Europe was easier than it has since become, many individuals who had received professional or scientific training migrated permanently from LDCs to the global core, attracted by income levels many times higher than they could ever expect to earn at home. Those who did return frequently carried with them cultural attitudes, professional practices, and salary expectations which aggravated rural–urban disparities. Although Seers argued that few of the countries most exposed to the loss of professional skills, such as India, appeared to have suffered serious damage, there are particular problems for smaller states. Sri Lanka, for instance, lost so many technicians to the Middle Eastern oil producers 'that it was forced to ask UNICEF to take over the National Water Board' (*Guardian Weekly* 26 April 1981).

Self-help in the periphery
Within the overall variety of the contemporary Third World, there are groups of LDCs which face similar development challenges. On many issues, states in each group share interests which conflict with, or at least differ from, those of core nations, so one might expect to find evidence of growing co-operation amongst them. In the jargon of the 1980s, this implies action on a 'South–South' basis, in contrast to the 'North–South' relationships we have just examined. However, a number of factors, many of them related to, if not actually a product of, the colonial experience, hamper such action. Hermassi (1980, p.172) argues that 'the central and crucial problem' facing Third World nations is that they 'are split societies in the economic, geographic, social, and political sense.' Sachs (1976, p.231) adds to this by suggesting that 'underdevelopment is characterized, and even defined, by the rigidity of structures, by the difficulty of getting out of ruts, by bottlenecks in the flow of goods.' We have noted the dendritic spatial structure, primate urban structure, and frequently the export-commodity industrial monostructure of Third World economies. Taken together, these characteristics point to some of the barriers which have impeded practical co-operation amongst LDCs.

The small size of so many Third World states, and their low levels of per capita GNP, undermine the potential benefits of a policy of import-substituting industrialization. Either plants are too small to achieve economies of scale, or else plants of the minimum economic size are grossly under-utilized. An obvious solution to this problem would be to promote regional common markets, with free trade among member countries, so that the efficiencies of large-scale production could be realized jointly (Pazos 1973). A number of such regional agreements exist, of which the Andean Pact has gone furthest to implement co-operative procedures for internal industrial specialization and the joint regulation of external investment in the region. However, the durability or effectiveness of many of these regional groupings has been limited. Apart from the political tensions amongst member states caused by divergent ideologies (which finally killed the East African Community of Kenya, Uganda, and Tanzania, for instance), there are invariably disputes about the spatial allocation of coordinated investments. Claims for an 'equitable' national share of industrial growth frequently clash with economic incentives to invest disproportionately in the regional core. Fragile nationalisms have a limited capacity to tolerate what appears as the tangible give, in contrast to the often intangible take, of effective regional co-operation. The same is true with respect to international commodity agreements amongst Third World producer countries. Although there are times when output restraints would be beneficial to all because of their price effects, there always seem to be some states tempted or constrained to increase their share of output, which undermines the rationale of collective action. The power of OPEC collapsed in the mid-1980s for this reason.

The larger, more industrialized LDCs, notably Brazil and India, are increasingly diversifying their exports of manufactured goods and professional services to other Third World states, although the growth of South–South trade faces a number of obstacles (Stewart 1976). Foreign direct investment on a South–South basis is also on the rise, especially in Asia. For instance, Indian firms accounted for nearly two-thirds of the total in Sri Lanka in 1979, and Hong Kong interests owned more than one-tenth of the total in both Indonesia and Malaysia (UN 1983a). Socialist critics might argue that far from such investment representing South–South co-operation, it merely adds another layer of dependency, and that its recipients are made doubly peripheral within the capitalist world-economy. The degree to which net benefits are likely to accrue to the host country will vary, for reasons already discussed. It is appropriate to note that Cuba and Vietnam, which have explicitly entered the orbit of the socialist bloc, still retain fundamentally dependent economic structures.

The greatest potential for 'self-help in the periphery' lies not at the international or national levels, but at the grassroots; in the local communities where the majority of the poorest citizens of Third World countries live. A long history of policies which discriminate against rural areas, and of cultural attitudes and social practices which discriminate against women, has shackled the development efforts of those who carry the most direct burden of agricultural production and maintenance of rural society. The emergence of local initiatives and institutions, particularly those which nourish the resourcefulness of rural women, has great promise, even if their initial achievements appear modest. Mackenzie (1986) documents the activities of informal women's groups in Central

Province, Kenya, which have both income-generating and communal welfare objectives. The former category are engaged in agricultural activities, handicraft production, or various commercial enterprises; the latter often act as savings and loan associations to facilitate household-related investments.

Conclusion

This chapter has established a basis for understanding both the heterogeneity of the Third World and its substantial commonality of experience *vis-à-vis* the core nations. By historical standards, the progress of economic development since 1945 has been impressive, yet serious problems remain; indeed, living standards have declined in many states during the 1980s. Priorities and practices need to change among both the ruling élites of most LDCs (see Ch. 5) and among the governments and multinational institutions of the world's industrialized states if the welfare of the world's poor majority is really to improve over the coming decades.

11 Challenges of Third World development

Agriculture and rural development

Introduction

With some obvious exceptions, such as Hong Kong, Singapore, and Kuwait, Third World nations face the same development challenge as that faced by the core states at an earlier date: how to transform a traditional agricultural economy into one which provides the basis for sustained economic growth and diversification. It is indicated in Table 11.1 that, although significant changes have taken place since the 1960s, lower-income LDCs especially remain fundamentally agrarian societies, heavily dependent on non-mineral primary exports for their foreign earnings and, as we shall see, in many cases struggling to achieve a precarious balance between the growth of agricultural output and of population. The prevailing wisdom of development specialists in the 1950s and 1960s encouraged newly independent LDCs to neglect what was seen as the 'stagnant' agricultural sector in favour of industrial expansion. Even since the failure of such policies has become apparent, the urbanized ruling élites of most Third World countries have shown little enthusiasm for giving the rural economy the priority and resources which its economic potential merits and its population deserves. The rural emphasis of the Chinese revolution (see Ch. 2) stands out as an exception which has been emulated less than it has been praised.

Because the thoroughgoing transformation of a traditional rural economy involves so many interdependencies, Mabogunje (1980) argues in favour of a 'big push' approach to agricultural development, and cites as examples the British enclosure movement, the homesteading of the Great Plains, the collectivization of Soviet and Chinese agriculture, and the creation of *ujamaa* village co-operatives in Tanzania. He suggests that without such coordinated and comprehensive action, attempts at piecemeal or incremental change are frustrated by rigidities elsewhere

Table 11.1   The declining significance of agriculture in national economies

	Labour force in agriculture, (%)		Agriculture's share of GDP, (%)		Non-mineral primary commodities' share of merchandise exports (%)	
	1960	1980	1960	1980	1960	1980
low-income economies	77	71	50	36	70	37
middle-income economies	61	44	24	15	59	27
industrial market economies	18	6	6	4	23	15
industrial command economies	41	16	21	15	33	11

Note: more recent data are not published on a comparable basis.
Source: compiled from World Bank 1983. *World development report 1983*; New York: Oxford University Press, Annex Tables 3, 9 & 19.

in the prevailing system. Both he and Chisholm (1982), who is critical of this argument, acknowledge that none of these examples, plucked out of their specific cultural and geographical contexts, can serve as models elsewhere. In the following sections we examine some of the elements of contemporary Third World agrarian economies, the interactions between which obviously vary considerably from country to country. Unless otherwise noted, data are derived from the survey of Third World agriculture in World Bank (1982).

Land
Rapid population growth has put increasing pressure on the land resource base in LDCs, to the extent that between 1961 and 1980 only one-fifth of the increase in agricultural production resulted from expansion of the area under cultivation, the remainder representing gains from intensification. There is some scope for further extension of agricultural settlement, but the potential is greatest in regions where existing levels of rural population density are least critical, notably in humid and subhumid parts of Latin America and Africa. China, in contrast, suffered a net loss of cultivated land during the 1960s and 1970s, and most of Asia is approaching its limits. Only by a generally successful record of increases in land productivity has the output of Third World agriculture kept ahead of population growth in recent decades, but the unevenness of this achievement, especially if one compares the progress made in Southeast Asia with the dramatic deterioration in Africa, is

indicated in Table 11.2. Divergent trends in population growth rates explain some of this difference, but the improved per capita output of Southeast Asian agriculture owes much to investments in irrigation and the increasing use of high-yield seeds and fertilizer. A more reliable and a more controlled water supply is the key to more intensive cultivation in large parts of the Third World, whether in the form of large-scale irrigation schemes (which are expensive and have often had adverse social and environmental side-effects) or of decentralized improvements in water and land management on the 80% of agricultural land which is rain-fed.

Tenure

The gradual rise in the productivity of British agriculture prior to the Industrial Revolution involved dramatic changes in the institutional structure of farming (see Ch. 4). The performance and prospects of contemporary Third World agriculture are similarly and intimately related to the social organization of access to land. The overwhelming majority of the Third World's rural population is poor: close to one billion people are desperately so. It is almost universally true that this rural poverty has been exacerbated by tenure systems which constrain attempts to raise long-term agricultural productivity. The improvement of rural living standards achieved in the wake of the Chinese revolution is but one indication that poverty is more than simply a function of a high population/ land ratio.

The major differences among continents of the Third World (together with comparative data for Europe and North America) in the territorial structure of agricultural land holdings are shown in Figure 11.1. In densely populated Asia, more than half of all farm units are less than 1ha in size, so that together they occupy only 10% of the farmed area. Units of 2–5 ha are the single most extensive category. Latin America, at the other extreme, has half of its farmed area concentrated in the top 2% of very large farm units (over 1000 ha) and, despite its lower population density than Asia, has over half of its farms crammed on to barely 2% of its agricultural land. Africa occupies something of an intermediate position, but one notes that over 90% of its farm holdings are less than 5 ha in size. A variety of social structures underlie these distributions. Overall, the rural poor (i.e. approximately 95% of the rural population of the Third World) can be divided into the 'more than half [who] are small farmers who own or lease their land; another 20 percent [who] are members of farming collectives, mainly in China; [and] the remaining one-fifth to one-quarter [who] are landless' and are overwhelmingly concentrated in the Indian subcontinent (World Bank 1982, p.78). Rich rural landowners are numerically insignificant but in

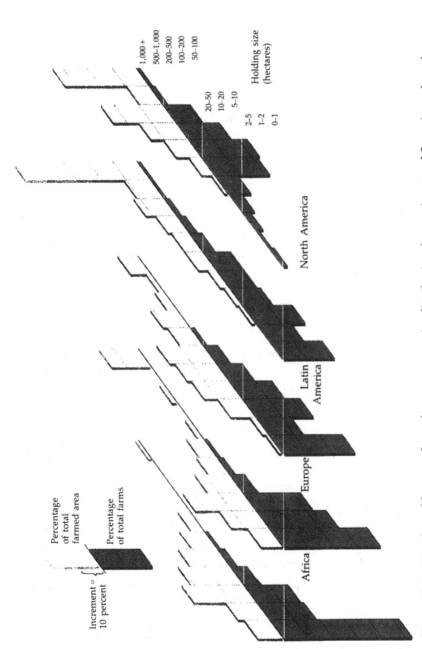

Figure 11.1 Number of farms vs. farmed area: comparative distribution, by continent and farm size, selected countries, 1970. (FAO *World Census of Agriculture* data)
Source: World Bank 1982. *World development report 1982.* New York: Oxford University Press; Figure 7.2.

Table 11.2   Growth rates of agricultural output by major world regions (excluding China). 1960–1980.

Region and country group	Agricultural output			
	Total		Per Capita	
	1960–70	1970–80	1960–70	1970–80
developing countries	2.8	2.7	0.3	0.3
low-income	2.5	2.1	0.2	−0.4
middle-income	2.9	3.1	0.4	0.7
Africa	2.7	1.3	0.2	−1.4
Middle East	2.5	2.7	0.0	0.0
Latin America	2.9	3.0	0.1	0.6
Southeast Asia	2.9	3.8	0.3	1.4
South Asia	2.5	2.2	0.1	0.0
southern Europe	3.1	3.5	1.8	1.9
industrial market economies	2.1	2.0	1.1	1.2
non-market industrial economies	3.2	1.7	2.2	0.9
world total	2.6	2.2	0.7	0.4

Source: World Bank 1982. *World development report 1982.* New York: Oxford University Press; Table 5.1

much of Latin America and the Indian subcontinent are politically and economically almost all-powerful, at least at the local scale which bounds the horizons of the poor.

Rural tenure systems in most of the Third World (excluding China) were profoundly affected by colonialism. The extreme concentration of landholding in Latin America originated with the imposition of a form of feudalism by the Spanish conquerors that evolved into a society dominated by an authoritarian and ultra-conservative 'landed baronial class' (Littwin 1974, p.35). In India, British rule generally consolidated a prevailing pattern of land ownership which was not notably egalitarian, and furthered the economic polarization of rural society by imposing monetary taxation, which increased the bondage of the poor to local élites. The treatment of land in Africa varied depending on whether or not extensive European settlement was contemplated (Davidson 1978). Where it was, as in South Africa, Kenya and Algeria, the usual practice was to expropriate the best land for settlers, which had the added 'advantage' of increasing the colony's supply of wage labour, as natives were displaced and impoverished. Where it was not, notably in British West Africa, taxation of the indigenous rural population aggravated conflict over rights to land in a continent marked by collective concepts of land use (Bates 1979).

It is against this background that the potential and prospects of agrarian reform in the contemporary Third World need to be judged. A detailed World Bank survey of one of the poorest rural areas in Latin America, northeastern Brazil, reveals some of the dimensions of what structural change would involve 'in a region where up to 10 million people remain near absolute poverty while agricultural land the size of France lies idle or greatly underused' (Kutcher & Scandizzo 1982, p.30). The distribution of landholdings reflects the continental pattern, with half the agricultural area occupied by 4% of all landowners, and one-third of all farms crowded on to 1.4% of the land. Half of the agricultural labour force has 'no formal or legal access to land and subsist[s] on temporary employment or scratch[es] out a living on landholdings so poor or so remote' that there is no official record of their existence (Kutcher & Scandizzo 1982, p.22). Farm size is the single most significant influence on the characteristics of agricultural production in the region, with the intensity of factor use declining rapidly as units increase in size (see Table 11.3). Simulation of the effect of various agricultural policies designed to improve the welfare of the rural population indicates that, *within the existing agrarian structure*, the best possible outcome would increase production by 20%, reduce consumer prices by 13%, and increase total employment less than 2%. But, if *land redistribution* were to take place on the basis of the farm modules proposed in 'the fully legislated, adequately funded, but officially unsupported' Brazilian Land Statute of 1964, the 37 000 largest farms (with an average size of 740 ha) could be converted into 790 000 family units of 35 ha. If such a reform were implemented, not only could agricultural output be expected to increase almost as much as under any of the non-structural policy options, but it 'would ensure employment and adequate consumption for more than twice the number of families the agricultural sector is currently accommodating' (Kutcher & Scandizzo 1982, pp.29–30). The greatest obstacle to achieving such change is that in societies where it could do most to improve the welfare of the rural poor, political and economic power resides in the landowning élite who are most threatened by it.

Incentives and infrastructure
Land reform is a necessary condition of rural economic development in much of the Third World, but it is not sufficient, by itself, to

Table 11.3   Farm Structure in north-east Brazil, 1974.

Farm size (ha)	Average holding (ha/farm)	Average land value (cr/ha)	Labour input (yr/ha)	Capital input (cr/ha)	Capital/ labour ratio (cr/worker)	Gross output (cr/ha)
0–9.9	4	1266	0.247	1203	4 870	769
10–49.9	27	1054	0.075	529	7 057	362
50–99.9	72	1064	0.041	405	9 872	291
100–199.9	141	1318	0.028	318	11 341	288
200–499.9	299	785	0.019	243	12 792	192
over 500	1180	625	0.012	140	11 625	121

Source: Kutcher, G. P. and P. L. Scandizzo 1982. The Agricultural Economy of Northeast Brazil. Baltimore: Published for the World Bank by The Johns Hopkins University Press.

guarantee greater prosperity. States with agrarian systems as diverse as those of Cuba (with its socialist-transformed plantation economy), Tanzania (with its collective *ujamaa* villages), Kenya, and India have discovered that stimulating the rural economy requires an integrated set of measures, the success of which hinges mainly on whether they support or stifle the initiative of the producer. Each currently existing socialist society, including China and the USSR, continues to grapple with defining the optimum balance between, on the one hand, securing the State's requirements with respect to agricultural output and, on the other, harnessing the enhanced motivation of producers who are allowed, at least to some extent, to work on their own account and sell surplus produce in unregulated markets. In market-economy LDCs, the challenge to the State is to avoid intervening in the agricultural sector in ways which discourage increased production. One such intervention, reflecting the urban bias of many Third World regimes, takes the form of policies that subsidize the cost of basic foodstuffs purchased by the (primarily urban) poor. Not only do these schemes frequently cost far more (as high as 20% of the national budget) than is allocated to investment in the rural economy, but they almost always depress the prices received by domestic farmers, reducing their incentive to expand production (World Bank 1982, 1986).

The greater productivity of small-scale Third World farms than of large landholdings is achieved despite many structural features of the agrarian economy which tend to favour the larger units. This situation frequently prevails even where state or para-state agencies are active, for their priorities often reflect either the interests of the landed rural élite (Griffin 1970a) or the self-interest of the agency's bureaucracy

(Ellis 1982). Access to credit at commercial rates is severely restricted to non-landowners, and small landowners are often deemed ineligible by financial institutions for lack of a clear legal title to their farm (a problem which is compounded in regions of traditional communal tenure). The majority of the rural population therefore depends on the 'unorganized' money market, where interest rates tend to be substantially higher than on institutional loans. Women, who are primarily responsible for staple food production in much of the Third World, suffer most directly from these forms of discrimination.

Investment in rural infrastructure, notably electrification and farm-to-market feeder roads, is vital if the welfare and economic diversification of the agrarian economy is to be advanced:

> In many parts of Africa, farmers are more than a day's walk from the nearest road. Measures to raise their productivity without transport and communications are of little use, because their crops cannot reach a market and they are cut off from new technology, inputs and ideas. Improved access usually brings with it an expansion of nonfarm enterprises such as shops, repair services, and grain mills (World Bank 1982, p.72).

Provision of transport services by native entrepreneurs is typically an important step in breaking down the isolation of rural regions, although McCall (1977, p.60) substantiates the concern commonly expressed by socialist writers about 'the potential for concentration [which] lies in the near-monopolistic position of rural transporters, and in how this power benefits a particular rural class.' Efforts to promote a greater awareness of improved production methods among small farmers through *agricultural extension services* are effective only if other infrastructural requirements are provided.

Output
With the arrival of Europeans, and subsequently North Americans, intent on establishing export-oriented agriculture, initially to supply tropical produce to temperate-region markets, the composition of Third World agricultural output began to change significantly (see p.71). Commercial monoculture of crops such as tea, coffee, sugar, bananas, palm oil, cotton, and rubber was most commonly introduced by the establishment of *plantations* (Courtenay 1980). These have generally occupied the best farmland, pre-empting its use to meet domestic food needs. The plantation economy has generally generated very few economic linkages with its host society in that the processing and/or marketing of the crop, as well as the supply of production inputs, is tightly controlled by the foreign owner. The characteristic labour-intensity of plantation agriculture has offered fewer benefits of employment-creation than

might be expected, partly because, in the past, labour was often imported by the company (a situation which guaranteed a dependent and poorly paid workforce), and partly because the routine tasks of cultivation and harvesting have in many cases been substantially mechanized. Graham & Floering (1984) argue, however, that the 'modern plantation' is very different from the traditional institution and contributes much more positively to host country development. Plantations established since 1960, especially in Southeast Asia, have transformed the world map of palm-oil production and have transferred skills and technology to the domestic economy in the process. Nevertheless, many states dependent on plantation exports for their foreign exchange remain vulnerable to unstable but generally stagnant earnings and are simultaneously forced to increase their imports of basic foodstuffs.

A new dimension of foreign involvement in Third World agriculture since the late 1960s is the growing output of products which are not specifically tropical. Improvements in global transportation and communication have made it possible for North American and Western European agribusiness corporations to exploit the cheaper land, cheaper labour, and longer growing season of tropical lands to displace or augment the supply of temperate region commodities from their traditional sources. The 'opening-up' of Amazonia, for example, has been associated with the creation of enormous beef ranches (holdings of 200 000–500 000 ha), generating very little employment (an average of one employee to 275 cattle), but yielding high profits (enhanced by Brazilian tax laws which promote frontier expansion). Investors in these operations include not only typical agribusiness firms, but also corporations such as Volkswagen and Mitsubishi (Garreau 1977, Hecht 1985). Feder (1980, p.220) cites an unpublished Brazilian government study which found that between 1964 and 1968 'over 32 million hectares were acquired by United States investors, over 10 percent of all farmland.' Lappé & Collins (1978) document a comparable form of agricultural 'development', involving the airfreighting of fresh vegetables from plantations in Senegal to European markets. Both these examples cast doubt on the benefits of schemes which increase the population/ resource ratio of LDCs by removing land from actual or potential domestic food production.

Progress in increasing the output of staple Third World food crops has been most obvious with respect to those favoured by the *Green Revolution*. As a result of genetic experimentation and development in international research centres devoted to rice, and to corn and wheat (located in the Philippines and in Mexico respectively), new strains of high-yielding grains began to be disseminated to Third World farmers in the mid-1960s. Their adoption brought about immediate and spectacular increases in output. In India, for instance, wheat production doubled between 1966 and 1971 and, following disease-related setbacks which had been combated by the mid-1970s, continued to rise in the 1980s to the extent that the country is now self-sufficient. The impact of high-yield varieties of rice in Southeast Asia lies behind that region's impressive increase in per capita food production, evident in Table 11.2.

The Green Revolution has generated a large body of literature embodying a wide variety of evaluations. Although the dangers of increasing the genetic uniformity of basic world food crops is one source of concern, the bulk of the controversy surrounding the introduction of high-yield varieties (HYVs) arises out of the sociopolitical implications of their adoption. The success of HYVs depends on the provision of a package of related requirements, among which chemical fertilizers and a reliable water supply are particularly significant. Richer (and more credit worthy) large landowners can more readily finance these inputs, and so were the first to adopt HYVs and benefit from their greater yields, whereas the first impact that many poorer farmers noted was a weakening of market prices as a result of the more abundant crops. The complexity of determining the net beneficiaries of the Green Revolution is explored by Lipton (1978). He points out that in many respects the introduction of HYVs benefits the poor: they have provided extra food, without which malnutrition and starvation would have been greater; and their greater sensitivity to careful and well timed cultivation calls for (and rewards) greater labour-intensity than traditional grain production. Nevertheless, for reasons we have just noted, HYVs are *institutionally* concentrated in larger farms, the owners of which exhibit a distinct tendency to promote mechanization (often with state-subsidized capital) at the expense of creating employment for the poor. At a continental scale, the poor farmers of Africa have so far benefited least from the Green Revolution, because research has neglected their staple crops.

Legitimate concern about the ability of a growing Third World population to feed itself adequately needs to be balanced by recognition of the considerable potential which still exists for food output to increase. Ladejinsky (1976) points to the outcome of national field demonstrations carried out in India in 1970–71, which indicated that, under optimum conditions, rice yields of four times the national average and wheat yields of three times the national average could be obtained. Even if only half of the potential increase could be realized, the implications for the welfare and economic development of Indian society would be profound. In India, as elsewhere, however, the institutional barriers to greater agricultural productivity noted above are often more difficult to overcome than the environmental ones.

The experience of West African oil producers has been particularly salutory in this respect (*The Economist* 5 December 1981). Prior to the commercial exploitation of oil, Nigeria and Gabon were among the strongest economies in Africa, their prosperity based on a steady growth in agricultural output and exports. Although the dynamic expansion of the oil industry through the 1970s doubled the annual rate of national economic growth, its side-effects were allowed to undermine the rural economy. Governments attempted to keep food prices low, resulting in a massive decline in marketed output, at the same time that high wages in the oil and related industries encouraged large-scale rural–urban migration. As a result, Gabon, which in 1964 produced twice its domestic food needs, was in 1980 forced to import 96% of its requirements. Nigeria also experienced a marked deterioration in its degree of agricultural self-sufficiency, importing $2.8 billion worth of food in 1980. The governments of Cameroon and the Ivory Coast, where the oil boom was just beginning in the early 1980s, appeared to take note and 'set farm prices high and tried to keep them there.' Indeed, 'the collapse of agriculture in Nigeria has handed [their] farmers . . . a big new market' (*The Economist* 5 December 1981, p.86). Not surprisingly, the lesson of the pioneering industrial nations, that sustained economic development necessitates the encouragement of a productive agricultural sector, is more widely acknowledged in LDCs today than it was in the 1950s, when rapid industrialization was seen as a short-cut to 'modernity' and rising prosperity.

Industrialization, urbanization, and regional disparities

The economic structure of Third World societies

The growth-oriented development strategies of the 1950s and 1960s embodied a conception of LDCs as *dual economies*. Although dualism was originally defined in terms of the ethnic and social cleavage evident in colonial societies between the native and the European elements of the population, it was soon adopted to characterize the (not unrelated) economic structure of these territories (Brookfield 1975). In simple terms, a basic distinction was drawn between the *traditional* sector and the *modern*, each of which was viewed as being essentially self-contained, both functionally and geographically. The traditional sector consisted of the indigenous economy, overwhelmingly rural, agricultural, employing very little capital, and only marginally oriented towards non-local markets. In contrast, the modern sector consisted almost entirely of European-controlled enterprises in the extractive (including plantation), manufacturing or service industries. These were relatively capital-intensive, market-oriented activities, concentrated in urban areas or in enclaves of 'modernity' in resource-rich peripheral regions. It was argued that a rapid expansion of the modern sector would allow LDCs to bypass the complex problems of increasing productivity in their traditional sector until such time as the dynamism of the modern

Table 11.4 Characteristics of the two circuits of the urban economy of underdeveloped countries.

Characteristics	Upper circuit	Lower circuit
technology	capital-intensive	labour-intensive
organization	bureaucratic	generally family-oriented
capital	abundant	scarce
hours of work	regular	irregular
regular wages	normal	not required
inventories	large quantities and/or high quality	small quantities; poor quality
prices	generally fixed	generally negotiable between buyer and seller
credit	banks and other institutions	personal, non-institutional
benefits	reduced to unity, but important due to the volume of business (except luxury items)	raised to unity, but small in relation to the volume of busines
relations with clientele	impersonal and/or through documents	direct; personal
fixed costs	important	negligible
publicity	necessary	none
re-use of goods	none; wasted	frequent
overhead capital	indispensable	not indispensable
government aid	important	none or almost none
direct dependence on foreign countries	great; outward–oriented activity	small or none

Source: Santos, M. 1977. Spatial dialectics: the two circuits of urban economy in underdeveloped countries. *Antipode* 9(3); Table 1.

sector itself made the task easier, notably by drawing increasing numbers of people into industrial employment. The 'stages' conceptualization of economic development (Fig. 10.1) suggested (rather superficially) that this had indeed been the sequence of events in many of the European core nations.

By the 1970s, however, the idea that the Third World was composed of dual economies was shown to be misleading, and policies based on it proved, not surprisingly, to yield disappointing results. Certainly, the economy of most LDCs seems to consist of two distinct sectors, but it is apparent that they are in fact, functionally interrelated, and that they cannot be viewed geographically in terms of a simple rural–urban dichotomy. The key to a better understanding of the structure of Third World economies is to give explicit recognition to their incorporation (with a few notable exceptions) into the capitalist world-economy. These two sectors are identified in Figure 11.2 as the *formal* and the *informal*, and some of the channels of interaction between them are indicated: comparable terminology refers to the *upper* and *lower circuits*, the characteristics of which are distinguished in Table 11.4.

Different authors focus on different specific variables to identify the source of this division. Santos (1977, p.50) argues that the upper circuit

> ... is a direct result of technological modernization, and its most representative elements are the [industrial and commercial] monopolies. Most of its relations take place outside the city and surrounding area, for this circuit has a national and international framework.

The lower circuit, in contrast, 'consists of small-scale activities and is especially concerned with the poor population.' As a result, whereas the lower circuit is present throughout the space-economy of a LDC, the upper circuit is concentrated in the metropolitan centres, and has a very limited institutional presence in rural areas. An alternative approach is to view the two sectors in terms of their labour force. McGee (1977a) draws on analyses of the *peasant* economy, which is essentially based on the full utilization of a family's labour supply, and thus points to the way in which traditional agricultural communities give rise to non-agricultural activities in the rural informal sector. The same form of organization begins to appear also in urban areas, where poor families unable to sustain themselves by wage employment in the formal sector seek to create jobs for themselves in the informal sector. Indeed, McGee suggests that Hart's analysis of income opportunities (Table 11.5) is particularly helpful in defining the two sectors.

Adopting the formal–informal classification of the domestic economies of non-socialist LDCs provides a framework for a better understanding of how their experience of industrialization and urbanization differs from that of the core nations in the 19th century. In particular, the superficial resemblance of contemporary phenomena such as rapid urbanization, large-scale rural–urban migration, and the emergence of a large service sector to key elements of European and North American development will be shown, in the following sections, to conceal a situation which holds more problems than promises for the peoples of the Third World.

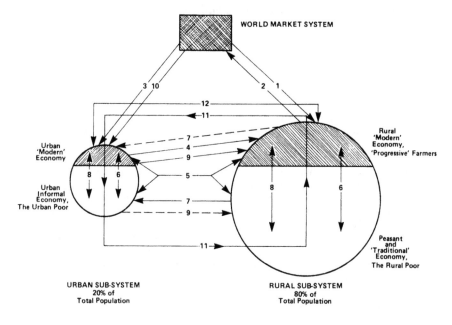

WORLD MARKET SYSTEM

Urban 'Modern' Economy

Urban Informal Economy, The Urban Poor

Rural 'Modern' Economy, 'Progressive' Farmers

Peasant and 'Traditional' Economy, The Rural Poor

URBAN SUB-SYSTEM
20% of
Total Population

RURAL SUB-SYSTEM
80% of
Total Population

TRANSFORMATION AND INTERACTION PROCESSES

1 The Colonial Impact: Land alienation and foreign investment in resource and commercial crop sectors.
2 Resource-based primary product export.
3 Foreign investment in import substituting industry.
4 Investment in rural development.
5 Increasing disparities between rural and urban sub-systems.
6 Increasing disparities between sub-sectors of both rural and urban economies.
7 Excessive Rural–Urban migration.
8 Patron–Client relationships.
9 Kinship-based economic remittances.
10 'Aid' flows.
11 Circular and return migration.
12 Indigenous capital investment.

STRUCTURAL CHARACTERISTICS

A. Government policies favouring a relatively open domestic market system.
B. Adequate resource endowments but limited availability of good arable land.
C. Relatively low technology.

Figure 11.2 A spatial sectoral model of the Kenyan economy.
Source: D. R. F. Taylor, Department of Geography, Carleton University, Ottawa.

Industrialization

An essential fact about Third World societies is that 'modern' industry, by which we mean the systematic harnessing of non-human energy to increase the productivity of the labourer which was at the heart of the Industrial Revolution, was implanted 'from the outside' by Europeans; it had not developed organically 'from within'. Moreover, from the beginning of the colonial era, core nations such as Spain and, pre-eminently, Britain adopted policies which actively retarded or suppressed the growth of manufacturing in their overseas empires (see Ch. 4). By

Table 11.5   Income opportunities in a Third World city.

Formal income opportunities
(a) Public-sector wages
(b) Private-sector wages
(c) Transfer payments – pensions, unemployment benefits

Informal income opportunities: legitimate
(a) Primary and secondary activities – farming, market gardening, building
     contractors and associated activities, self-employed artisans, shoemakers,
     tailors, manufacturers of beer and spirits
(b) Tertiary enterprises with relatively large capital inputs – housing, transport,
     utilities, commodity speculation, rentier activities
(c) Small-scale distribution – market operatives, petty traders, street hawkers,
     caterers in food and drink, bar attendants, carriers (*kayakaya*), commission
     agents, and dealers
(d) Other services – musicians, launderers, shoeshiners, bankers, night-soil
     removers, photographers, vehicle repair and maintenance workers;
     brokerage and middlemanship (the *maigida* system in markets, law courts,
     etc.); ritual services, magic, and medicine
(e) Private transfer payments – gifts and similar flows of money and goods
     between persons; borrowing, begging

Informal income opportunities: illegitimate
(a) Services – hustlers and spivs in general; receivers of stolen goods; usury, and
     pawnbroking (at illegal interest rates); drug-pushing, prostitution, poncing
     ('pilot boy'), smuggling, bribery, political corruption Tammany Hall-style,
     protection rackets
(b) Transfers – petty theft (e.g. pickpockets), larceny (e.g. burglary and armed
     robbery), speculation and embezzlement, confidence tricksters (e.g. money
     doublers), gambling

Source: Hart, K. 1973. Informal income opportunities and urban employment in Ghana.
Journal of Modern African Studies 11 p69.

---

the time that Latin America achieved its political independence, Great
Britain was well on the way to becoming 'the workshop of the world',
and when decolonization came to Africa and Asia in the middle of the
20th century, the typical large US industrial corporation was on the verge
of becoming truly multinational. Modern, formal-sector, industry in the
Third World, therefore, is disproportionately small in amount, and either
foreign-controlled or, as sovereign governments have sought to remedy
these two perceived weaknesses, state-owned. In 1980 the Third World
contained 65% of the world's population but produced only 10.9% of
the value added in world manufacturing production (UNIDO 1981).

Although most writers and governments are referring to formal-
sector industry when they talk about industrialization (and when they
collect statistical information), approximately 70% of the industrial

workforce in the Third World is employed in small-scale enterprises with informal-sector characteristics (Norcliffe 1982). The majority of these are located in rural regions, reflecting the numerical dominance of the rural population, and are engaged in the production of cloth, handicrafts, simple tools, and repair work. The further development of rural industry depends ultimately on the increased prosperity of the agricultural economy, and calls for many of the same policies to promote it, such as improved provision of infrastructure and credit. It also focuses the need for an explicit strategy of rural central-place development, so that beneficial agglomeration and linkage effects may be realized (UNIDO 1979).

The range of output and of linkages with formal-sector producers is greater among urban enterprises than their rural counterparts. Some *petty producers* in the urban informal manufacturing sector act as subcontractors to formal-sector firms, accepting work on a putting-out basis, for instance in the garment business. The majority, however, are primarily occupied in meeting the needs of the poor:

> They are not making 'the same product' as the big firms, . . . they are producing an inferior and cheaper product in a way which makes it accessible to persons who can neither buy in large quantities nor transport for long distances. This principle is equally true for bricks and groceries (Peattie 1980, p.25).

Nevertheless, over time, there is a tendency for formal-sector manufacturers to invade the markets of petty producers, for instance with mass-produced cooking pots or plastic sandals, which displace traditional products. The informal sector thus suffers a gradual contraction in the scope of its output and finds it increasingly difficult to accumulate the capital which would enable it to invest in improved technologies (Obregon 1974). Activities which are not directly threatened by the expansion of the formal sector may still be very reliant on it, in various ways. In a study of informal sector linkages in Dakar, Gerry (1979) noted both significant volumes of purchases by petty producers of material or tools from the formal sector, and also a considerable history of wage employment through which skills and capital had been acquired.

Formal-sector, or 'modern' manufacturing is overwhelmingly concentrated in urban areas, and frequently in just one or two metropolitan centres. Mabogunje (1973) notes, for instance, the industrial prominence of the 'capital-cum-port cities' of tropical Africa, which combine the advantages of their strategic location for assembling or processing imported goods with being the national focus of institutional spending and of consumers with high disposable incomes. The absence of appropriate industrial infrastructure (transportation, electric power, water supply, etc.) is a major constraint on dispersing manufacturing to smaller urban centres, although institutional and attitudinal factors are probably even more decisive in maintaining the metropolitan concentration of industry in the face of dramatic increases in various congestion costs, not least of traffic (Ayeni 1981).

The structure of formal-sector manufacturing in the Third World differs significantly from that in the industrialized nations. The two largest sectors in LDCs in 1975 were textiles and food products, which together accounted for 39% of manufacturing employment, more than twice their share in the Western industrial nations and Comecon. In contrast, the non-electrical machinery and transport equipment sectors, the leading employers in these industrialized nations, were relatively only half as large in the LDCs, where they together provided 10% of manufacturing jobs. During the 1970s, the two largest sectors grew more slowly than sectors producing capital goods (UNIDO 1981). However, we need to remember that not only does the Third World account for a small proportion of the world's modern industry (approximately 30% of employment, but only 11% of value added), but this capacity is highly concentrated in no more than about ten nations, of which Brazil, Mexico, India, Argentina, South Korea, and Taiwan are clearly the most prominent in terms of value-added (World Bank 1982).

Some of the characteristics of Third World industrialization have already been touched upon. Initially, most LDCs focused on the provision of basic consumer goods, such as textiles, for the domestic market, or on the processing of local export-oriented agricultural produce. Many of the currently more industrialized states experienced a considerable growth and diversification of their manufacturing sector between 1914 and 1945, when their links with Europe were disrupted by two world wars and the Depression. (The wars increased their primary product earnings but reduced the flow of European manufactures; the Depression so reduced their export earnings that they could not afford to buy European goods: both situations promoted domestic manufacturing.) During this phase, production was concentrated in the hands of European firms, or, in sectors with a simple, well established technology, of firms owned by the domestic economic élite.

The faster pace of industrialization which the Third World has exhibited since the 1950s has been characterized by the widespread penetration of TNCs and the substantial involvement of the State, both in the newly independent LDCs and in Latin America. Industrial policies have reflected differences in the fundamental ideological orientation of states (Fig. 5.2). Almost every Third World country has encouraged the growth of import-substituting industry by granting it some form of tariff protection, but only a few nations have avoided the dangers of industrial stagnation to which that approach readily leads. The most dynamic manufacturing economies, the NICs, are those which have gone on to foster export-oriented production which is fully competitive in global markets.

A policy of import substitution can be as valid today as it was when advocated by states such as Germany and the USA in the 19th century. In the face of hegemonic economic power in world markets, it is politically desirable and economically rational for a government to protect domestic *infant industries* which, once established (i.e. having overcome the prevailing barriers to entry; see Ch. 6), are potentially capable of producing as efficiently as their foreign competitors, at which point they no longer need protection. There are gains to employment and to the national balance of payments in this case, but it requires a well informed and competent government to protect the appropriate industries, to prevent them from being taken over by TNCs, and to phase out the protection at an appropriate time. These conditions are rarely present in LDCs where TNCs are dealing with 'soft states' (Weinstein 1976).

Nations ruled by conservative, nationalistic élites, and exhibiting the pronounced income disparities associated with this ideological orientation, protect the wrong industries and are politically incapable of removing the protection, for supporters of the regime are its principal beneficiaries. The political and economic interests of these *comprador élites* are more strongly tied to the interests of the foreign investors, whose activities they facilitate and legitimate, than to those of the poor majority of their fellow citizens. Instead of encouraging industry which has the potential to achieve economies of scale because its output is geared to meeting basic domestic requirements for items such as soap or cement (and note that the small size of many LDC markets is a handicap in this context), the greatest incentives are given to producers of 'non-essential' consumer goods, for which the affluent metropolitan élite is the only market and for which core-region TNCs are the predominant source. The result is a manufacturing sector which exhibits to an extreme degree the 'miniature replica' problem (see Ch. 7), characterized by relatively capital-intensive but grossly under-utilized plants, selling high-cost products in a profitably protected market which soon becomes saturated because living standards among the mass of the population remain depressed. This form of industrialization, which has typified much of Latin America, creates little employment and tends to have a high import-propensity (both for components and capital equipment) which negates its theoretical balance-of-payments effect (Colman & Nixson 1978). Insofar as this pattern of output is maintained by regimes which displace the TNCs by domestic private-sector firms or para-statal industrial enterprises, the implications for national development remain unaltered. It is noticeable, however, that the weak economic growth and rising indebtedness of many Third World nations in the 1980s has led to a much greater interest in stimulating efficient production, evidenced by widespread privatization of state-owned firms and a warmer welcome to investment by TNCs.

Despite their low levels of economic development compared to the industrialized core nations, Third World states such as Brazil, India, and China have the diversified resource base, large domestic market and, increasingly, the necessary physical and educational infrastructure to support a substantial manufacturing sector capable of making technological progress independently of (although certainly assisted by) the resources of US, European, or Japanese TNCs. However, if most of the smaller LDCs are to exploit the potential of a modern manufacturing sector to create wealth and employment, they can only do so by capturing foreign markets. This requires an outward-looking policy, and hence either a commitment to relatively free trade within the 'new international division of labour' (see Ch. 7) or else finding an uncertain niche within a socialist international division of labour which has yet to foster any significant growth of manufacturing in the Third World. The former strategy, being followed most explicitly by the Asian NICs, may indeed represent a more 'dependent' mode of industrialization than that of the 19th-century pioneers, but it is one which appears to offer the greatest potential of currently available options. Japan's success in growing to industrial maturity by this route is no guarantee that other states, in a different era of 'world time', will repeat its performance. Nor, given the difficulty which Western core nations are having in restructuring their domestic economies to accommodate increasing volumes of manufactured imports from the NICs, is it likely that many other LDCs will be able to emulate Taiwan and South Korea in the near future (Bradford 1982).

Keesing (1979, p.152) notes bluntly the domestic political correlates of rapid economic growth based on manufactured exports:

> the roster of governments that have succeeded in instituting coherent, pro-trade policies is not exactly a list of repression-free, anti-authoritarian regimes . . .. Having labor leaders in jail as well as achieving enough political stability to take a long-run view seem to be very common accompaniments.

He goes on to suggest, however, that the success of these policies is more likely to increase social harmony than social polarization. Bienefeld (1981, p.87) confirms that, in the case of Taiwan and South Korea at least, 'capitalist development has proceeded to the point of beginning to diffuse benefits and pushing up unskilled real wages.' (Significantly, these are also nations which have supported, rather than discriminated against, their farm sector.) Does this imply than an authoritarian state which is 'progressive' (defined in either socialist or capitalist terms) rather than conservative is, in fact, a necessary condition for Third World development in the contemporary core-dominated global economy, because sustained development requires policies with short-term effects that are 'unlikely to be acceptable on any democratic basis' (Bienefeld 1981, p.93)? There is no simple answer to such a question, but it may point to one of the costs faced by LDCs if they are to build a dynamic and diversified industrial sector in the contemporary era of 'world time'. One thing is certain: the trade policies of core nations have a significant impact on the prospects for Third World manufacturing.

Urbanization
With the onset of the Industrial Revoution, urbanization in the core regions of the world became associated, above all else, with the growth of manufacturing. The service economy which now dominates the metropolises of North America and Europe is to a large extent an outgrowth of the goods-producing system and of the personal needs of the urban population. In contrast, manufacturing has played a minimal rôle in the evolution of the Third World city. Urban centres developed by colonial authorities were, above all, places of administrative control (civil and military) and the regulation of trade. In territories where the indigenous society had developed its own urban centres, their functions resembled those of the pre-industrial European city (see Ch.4). The imperial powers tended either to ignore these settlements or, in strategic locations, to establish an adjacent but separate urban centre, so that the distinction between the upper and lower circuits of the urban economy and its social institutions was given a clear morphological expression.

The achievement of political independence has brought little change to most Third World metropolises, in the sense that formally, functionally, and attitudinally the colonial élite has been replaced by a domestic one. The poor are not wanted in the city; so the informal sector is constricted or harassed, and squatters, denied the security of having their presence legalized, are sporadically uprooted (Abrams 1977). Meanwhile, the inherited spatial form (specific reference here is to India) places

> . . . immense strains on the economic resources of the post-colonial city. Because colonial administrators wanted 'views', status, privacy and a particular cultural environment, water pipes, sewers, telephone wires, cables, roads and transport routes have to be perhaps five times as long as they need be . . .; low-income workers . . . have to travel across acres of 'colonial space' in order to reach centrally located work-places . . ., [their] productivity . . . affected not only by the hours lost in travelling but also by the sheer physical strain of waiting and travelling in [tropical] climatic conditions (King 1976, pp.284–5).

The scale of urban population growth in the LDCs since 1950 has been such, however, that neither policies designed to improve the welfare of the poor nor those designed to deter them from taking root in the urban area have significantly altered the dynamics of urbanization.

The world's urban population more than tripled between 1950 and 1986, increasing from 600 million to two billion people. In absolute terms, most of the growth took place in the Third World, which now contains over half the world's urbanized inhabitants. Although the level of urbanization varies significantly between continents, from 24% in South Asia to 65% in Latin America and Oceania (1986 data), all parts of the Third World have experienced this rapid growth (Brown & Jacobson 1987). Natural increase, in societies which are in the rapidly expanding stages of the demographic transition, accounts for approximately half of it. The balance represents the effect of large-scale rural–urban migration from regions experiencing increasing population pressure on their resource base, aggravated by structural problems in the agrarian economy of the sort described above (Drakakis-Smith 1981).

The nature of this migration is quite complex (McGee 1977b). *Permanent* residence in a metropolis may be preceded by a period of seasonal transfers between the village and the city or, in the relatively urbanized context of Latin America, by a step pattern of migration from rural areas through a regional central-place hierarchy (Nelson 1969). *Periodic* migration between the agrarian economy and formal-sector employment in the city or a mining camp is widespread, but especially significant in southern Africa (Lemon 1982). *Circulatory* migration is particularly well developed in Southeast Asia (McGee 1981). Residents of a particular village may establish an informal-sector enterprise, such as a bakery or an ice-cream vending operation, based on one or more group houses located in the metropolis. A group of villagers will run the business for a period, before returning to their rural 'home' while another set of villagers leave for the city to replace them. Improved transportation facilities have led to an increase in the volume and geographical radius of such circulatory movements.

Despite the poor living conditions and low-income employment opportunities offered by the squatter settlements of large Third World cities, their inhabitants compare life in them favourably with the options they would have faced had they stayed in the rural economy. Friedmann & Wulff (1979, p.29) summarize a number of studies which suggest that the better prospects of making a living in urban areas, together with 'the abysmal neglect suffered by rural areas at the hands of the governing authorities' in many LDCs (evident in the poor quality of health and education provision and often the lack of security from violence) are decisive in stimulating rural–urban migration. Nelson's study (1969) of migrants in Brazil suggested that new arrivals found urban jobs fairly quickly, thanks to the kinship or community networks which link the informal sector to the agrarian economy, and that eventually they equal or improve upon their previous socio-economic standing. More recent experience is less encouraging. The revolutionary potential of a poor, disenchanted urban proletariat, which many writers expect to develop in the squatter settlements of the Third World, has begun to materialize. Despite the apparently huge capacity of the informal sector of the urban economy to absorb labour in a wide variety of occupations (Table 11.5), there are limits to its ability to support a perpetually growing population at a constant standard of living. In other words, a process of *urban involution*, with dynamics similar to the agricultural involution portrayed in Figure 2.3, could well accelerate in countries in which the overall level of economic development stagnates, and whose governments do little to improve the prospects of the rural masses (McGee 1971).

Regional disparities and regional development
The relative poverty of Third World nations is accompanied by inequalities in their domestic income distribution, both between persons and between geographical regions. These are in most cases significantly greater than those found in the Western industrialized nations (Gore 1984). In societies which are still predominantly agrarian, the highly concentrated ownership of land (Fig. 11.1) is a major source of interpersonal inequality, as is the fact that well paying jobs in the formal sector of the economy or in the higher levels of government administration employ only a few per cent of the national population. Regional income disparities are directly associated with interpersonal disparities because of the overwhelming concentration of well paying jobs in one or two metropolitan centres, and the preference of landed élites to reside there also, close to the levers of power and the material basis of 'good living'. The almost total absence, outside of socialist LDCs, of vigorous policies to combat interpersonal and interregional economic disparity points to the profoundly different sociopolitical structure of most Third World societies, as compared to the mixed-capitalist democracies of Europe and North America – much more than to their relative lack of resources to implement redistributive measures. The fact which stands out clearly is that whether or not a state adopts an explicit regional development policy, the institutional forms of its national development strategy will profoundly influence the geographical distribution of economic activity.

As noted in Chapter 10, distribution issues were played down by definitions of development which equated it with achieving a rapid rate of economic growth. Kuznets (1955) suggested reasons why disparities in income distribution might be expected to worsen in the early phase of the development process, but then to decline as the benefits of growth diffused to a growing proportion of the population, leading to an 'inverted U-shaped curve' of disparity plotted against per capita GNP. Williamson (1965) pursued the spatial implications of this hypothesis, and assembled data which he argued confirmed that regional income inequalities were increasing in LDCs, but that they had declined, relatively, in developed nations, and hence were likely to follow suit in LDCs eventually. Gilbert & Goodman (1976, p.119) present more recent findings which indicate, however, that Williamson's conclusions cannot be regarded as definitive, for 'no consistent association between levels of regional income inequality and economic development can be detected.'

The diffusionist concept of development underlying Williamson's study treats the State authorities as neutral servants of 'the public interest' and so ignores the influence of the political ideology of ruling Third World élites, whose urban and anti-egalitarian bias is strong. Rogerson (1980) suggests a modification of the 'inverted-U' hypothesis which takes the nature of the State explicitly into account (Fig. 11.3). Nations ruled by a conservative, nationalist élite (the lower right quadrant of Fig. 5.2) are likely to experience a pronounced and continuing divergence between the prosperity of the national metropolis and the neglected rural periphery. States, such as India, which effectively regulate the channels of foreign economic penetration and attempt to foster a relatively autonomous and more egalitarian pattern of development are most likely to conform to the Williamson model. If pronounced interregional (and interpersonal) disparities promote

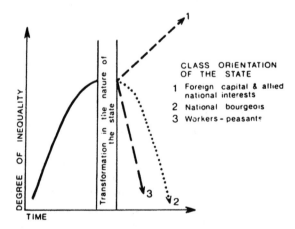

Figure 11.3   The trajectory of income disparities during the course of economic development in relation to the national political economy.
Source: Rogerson, C. M. 1980. Internal colonialism, transnationalization and spatial inequality. *South African Geographical Journal* 62; figure 5.

a socialist revolution, as in Cuba, then Rogerson suggests that the convergence of income levels across the country will be accelerated. In practice, the speed and degree of convergence will depend on the size of the country (given the limited resources of LDCs), and the success of the authorities in checking rural–urban migration. The anti-urban ideology of many socialist regimes promotes lower levels of urbanization than would prevail in a capitalist state at a comparable stage of development (Murray & Szelenyi 1984; Fig. 11.4). Slater (1982) critically assesses Cuba's achievements in this sphere, and Murphey (1980) provides a comparative evaluation of Chinese and Indian performance.

To counter the urban-focused, 'top-down' approaches to development, which tend to improve the income levels of a limited group of people who find formal-sector employment but which have very restricted spread effects, a number of writers have called for 'bottom-up' strategies (Stohr & Taylor 1981). These have an explicit rural bias and promote, as a major objective, a high degree of local or regional self-reliance both materially and institutionally (administratively). Friedmann & Weaver (1979), specifically addressing the possibilities in densely populated rural areas such as characterize India and China, invoke the concept of *agropolitan* development. This calls for the 'selective spatial closure' of a regional economy around urbanized centres which function as integral parts of the agrarian system

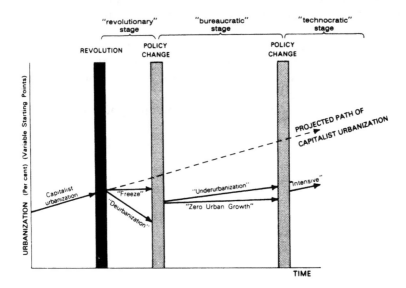

Figure 11.4   A model of socialist urbanization (after Murray & Szelenyi 1984).
Source: Thrift, N. & D. Forbes 1986. *The price of war: urbanization in Vietnam 1954–1985.* London: Allen & Unwin; Figure 3.1.

rather than as dissociated foci of external dominance. Insofar as land reform is a precondition for development strategies which encourage greater local autonomy, a strong central authority that can implement the necessary changes is also implied. However, as Stohr & Taylor (1981, p.471) observe, 'national power groups normally ... strongly resist giving up power once they have acquired it.' Nevertheless, some states, varying substantially in size and political philosophy (e.g. China and South Korea), do appear to have followed development policies which are relatively favourable to the rural poor.

Conclusion

The diversity of resources, cultural and political values, inherited economic orientation and recent development experience across the Third World makes generalization dangerous. This chapter has focused on key dimensions of national economic life and attempted to indicate the sort of issues that they raise. Solutions to each country's development challenges can only be adequately addressed in their immediate context. A minority (but a growing one) of Third World heads of state have acknowledged that many of the problems and privations that their people face will not improve without a disciplined reappraisal of indigenous public life. This necessary step needs to be matched, however, by a similar reappraisal, by core governments and the institutions that they control or influence, of their commitment to promoting a more prosperous world. The Third World debt crisis is partly the result of foolish behaviour by First World banks as well as by Third World governments (George 1988). Is there the will to make solutions a joint venture of core and periphery also?

12 *Prospects for the global economy*

Global interdependencies

Societies in every part of the world face serious challenges in the closing decade of the 20th century. The industrialized capitalist nations are confronted with the economic and political stresses of adjustment to an increasingly volatile international system in which their power to influence events (at home as well as abroad) has diminished. The industrialized state-socialist nations are even harder pressed. The economic achievements of a centralized, command economy are widely recognized to have reached their limits; yet the transition to a more dynamic and flexible economy involves a devolution of decision-making that is alien to traditional Soviet-style regimes. Third World nationals now look back on at least a generation of political independence, and most wonder what has happened to the high hopes of economic development that accompanied the achievement of statehood. Poverty, hunger, armed conflict, and political disillusionment remain widespread. There have undoubtedly been significant advances in freedom from want and oppression in some countries, but the overall picture is not exactly glowing. There is, finally, an increasing global awareness that the environmental consequences of economic growth (and of continued underdevelopment) have been too long neglected and now call for urgent attention.

The genuinely *interdependent* character of the global economic system is gradually becoming more widely recognized, and that is a positive step. In theory at least, people acknowledge that the problems which we currently face (of job-creation here, of more sustainable resource utilization there, or of creating more constructive mechanisms of international development) cannot be treated in isolation. Brazil's indebtedness, for instance, undermines employment in the USA and threatens the climate of the entire globe. Such linkages are spelled out more precisely below. But policies which address the implications of global interdependence effectively are disappointingly slow to take shape. Situations which are perceived to threaten the immediate interests of core states still meet too readily with a retreat into neomercantilism, as richer nations attempt to insulate their domestic economies from 'disruptive' change in the wider world, and thereby accentuate its social and environmental costs elsewhere. Are more creative responses to the challenges of an interdependent world discernable?

Three issues of particular importance at the close of the 20th century pose this question more concretely. If the purpose of economic activity is ultimately to promote human welfare, has this goal been lost sight of? The desire for peace, sustainable development of the natural environment and opportunity to engage in productive work is frustrated in the daily experience of millions of men and women in a world where the arms race, wisespread under- and unemployment, and signs of environmental degradation loom large. The following sections demonstrate some links between the satisfaction of these aspirations and the future evolution of the global economic system.

The implications of militarism

The capacity to threaten institutionalized violence has been a cornerstone of political power throughout history. The economic consequences of maintaining a permanent military establishment and, even more, of using it to wage war (or suppress internal opposition) have been profound. The cost has been a major factor behind the growth of taxation, and also of inflation (Hirsch 1978). The innovation and perfecting of armaments has been a consistent interest of the State (see p.86), especially in nations aspiring to, or seeking to maintain, hegemony. In this century, government funding of R&D with military applications has significantly shaped the pattern of civilian industrial innovation. To the direct costs of maintaining armed forces must be added the opportunity costs of consuming vast resources for this purpose rather than putting them to civilian use, whether in the form of public services or of directly productive investment (Stein 1985). Adequate account also needs to be taken of the direct and indirect costs of warfare inflicted on the environment and resident population of territory caught up in hostilities (Hewitt 1983b). Since 1945, these have been borne almost entirely by inhabitants of the Third World. All told, the impact of militarism on the global economic system is enormous.

Growing sensitivity to its implications stems from a number of distinct but related considerations. Undoubtedly, the horrific destructive potential of nuclear weaponry, production of which has long passed all (even strategic) *reason* (Myrdal 1978), has impinged on the consciousness of 'ordinary' men and women in many parts of the world. Sophisticated geographical modelling of the likely outcome of a nuclear attack on the UK (Openshaw, Steadman & Greene 1983; Openshaw & Steadman 1983a, b) has demonstrated not only consistent official bias towards underestimating the number of casualties, but has also pinpointed the massive destruction of national productive capacity and economic infrastructure that would ensue. If opposition to nuclear arms appears to be most deeply felt in Western Europe, there is good reason, for the logic of nuclear conflict between the superpowers has flowed from the anticipation of military confrontation in that continent. Yet the consequences of a nuclear war would be much more widespread and thoroughly devastating. The concept of a *nuclear winter*, which has been carefully refined since the early 1980s, suggests that the smoke and dust resulting from the detonation of warheads would, if 'spread over the mid-latitudes of the Northern Hemisphere, [reduce] the sunlight reaching the ground ... by least 90 percent ... [D]isruptions in agricultural productivity and/or food trade and aid' would lead fairly rapidly to massive starvation (World Resources Institute 1986, p.178).

A second source of concern is the global proliferation of military arsenals, primarily of 'conventional' weapons, to nations and dissident factions that are less politically stable than the superpowers. The widespread availability, and certainly the use, of increasingly sophisticated armaments directly undermines prospects for the peaceful resolution of conflicts between nations. It could be argued that to deplore the acquisition by nations of the periphery of military equipment which core nations possess in vastly greater quantities reflects a cynical ideology designed to maintain the core's dominance. But one notes that this equipment is deployed primarily in conflicts which take place within or between states of the periphery. And this feature highlights the third source of concern; the cost of the arms race, its distortion of resource allocation in domestic economies, and the opportunity cost of arms expenditures in a world which could deploy those resources so much more beneficially (Sivard 1986).

By the mid-1980s, global military expenditures were running at an annual level of over $900 billion. Although this represented 'only' about 5.5% of 'world GNP', the rate of spending had accelerated since 1973, whereas rates of economic growth had declined. The USA and the USSR are together responsible for 60% of the outlays, but the most rapidly increasingly share of expenditures since the early 1960s has been that of the Third World, rising from 8 to 20%. 'Arms imports of developing countries between 1975 and 1985 amounted to 40 percent of the increase of their foreign debt in that period' (Sivard 1986, p.5). To put these sums in perspective, Sivard contrasts military with social expenditures worldwide. In 1983, per capita spending on military forces, education, and health care was $523, $490, and $454 respectively in the developed economies of East and West, and $40, $27, and $11 respectively in the LDCs. In other words, the Third World as a whole spends more on armaments and military personnel than it invests in the training and physical wellbeing of its citizens. In 1980, for every $100 spent on military forces, less than six cents were spent on international peacekeeping. And in a world whose economic disparities between core and periphery are a major source of tension, the net economic aid given by Western core nations to poorer countries was 8% of their military outlays. Soviet aid was 1%.

It is important to recognize that military interests (and budgets) have become institutionalized in the contemporary world in ways which make the reallocation of resources to more constructive uses much more difficult to achieve than it was in the past. It was a soldier-president (Eisenhower), not a radical intellectual, who issued 'the first serious warning about the "military–industrial complex"' which took shape as a novel phenomenon in the USA in the aftermath of the Second World War (Sampson 1977, p.102). The origin of state-socialist societies in the successful use of revolutionary force has, most obviously in the USSR, given military interests a privileged position in their ruling councils and mechanisms of resource allocation. Military-dominated governments have also, for a variety of reasons (Horowitz 1982), proliferated among Third World states in the decades since their independence; so that in the global periphery, as in its core, economic decision-making is heavily influenced by the perceived needs of the military establishment.

Research and development to produce ever more 'complex, accurate, and lethal instruments of war' (Sivard 1982, p.9) dominates public expenditure on technological innovation in the global core. The proportion of military spending devoted to R&D has risen from around 1% before the Second World War to between 10 and 15% today. The continuous demand for 'new, improved' weapons systems, without which national security will be 'critically jeopardized', results in escalating expenditures on equipment the technological obsolescence of which follows quickly with relentless logic. It is hardly surprising that some genuinely productive spin-offs are generated by a research effort of this magnitude, but neither is it accidental that, among industrialized Western nations, the annual rate of productivity growth in manufacturing between 1964 and 1984 was strongly and inversely correlated with the proportion of GNP devoted to military expenditures (Sivard 1986). The absence of a negative correlation between arms spending and economic growth in the Third World, overall, is explained by Luckham (1978, p.44) on the basis that 'resources for armaments are typically diverted from social welfare spending rather than from productive investment.' This is a choice which 'most probably can only be [made] by governments which are prepared in the final analysis to repress the discontent it brings about.'

The technological intensity of modern armaments has noticeably widened the core–periphery gap in weapons production capability. A number of countries outside the global core, such as Israel, Brazil, and South Africa, have developed a significant indigenous armaments industry, but the global proliferation of weaponry has been overwhelmingly associated with arms exports from the core states of East and West. In Western Europe more so than in the US, the pressure to promote foreign arms sales to help offset the financial burden of maintaining a state-of-the-art production system is strong. The willingness of core nations to transfer high-technology products in the military field exhibits a rapidity which is significantly greater than that evident in civilian applications. The USSR, the USA, and Western European nations alike responded to their increasing financial difficulties during the 1970s by stepping up their international arms sales. A substantial volume of OPEC petrodollars was recycled in this way.

The employment consequences of militarism are also pervasive. The mobilization of unemployed individuals by authoritarian régimes both increases the government's power and reduces a potential source of civil unrest. Hitler seized on this fact in Germany during the Depression, and similar considerations have certainly prompted the build-up of the armed forces in many contemporary Third World states. In Western core nations it is employment in the military supply industries which tends to exert greater influence on policy-makers (Markusen 1986). Regional concentrations, such as that in California, where the largest employers are defence contractors, provide the basis for political constituencies supportive of continued arms spending (Sampson 1977). The desire to maintain jobs in high-technology sectors such as military aerospace has justified government subsidies, or takeovers, of firms in Britain and France.

To say that we would all be better off if the human and material resources currently devoted to military purposes were released for economic and social development, especially in the Third World,

> . . . is altogether too simple. It would be fortuitous if the structural shifts out of armaments production were exactly to match the new requirements for basic needs production in terms of either employment, investment, exports or imports, public revenue or expenditure – let alone all of these together with the political repercussions which they would set in train (Jolly 1978, p.105).

Nevertheless, even in a country as rich as the USA, current levels of military spending can no longer be ignored: they raise fundamental questions about the economic and political goals to which the nation aspires, and about the configuration of a world in which it seeks to enjoy peace and security. The USSR faces these choices even more acutely.

The economics of environmental change

Although the cataclysmic interpretations of the original *Limits to growth* report (Meadows *et al.* 1972) have been discredited, there is a consensus that changes in environmental conditions are critical to the prospects for improved human welfare in the coming decades. This conclusion applies both to the Third World and to more affluent societies, and it is valid (although not unaffected by) whatever growth rates their various national economies manage to attain (US Government 1980). The links between economic activity and the behaviour of environmental systems are increasingly seen to be two-way. Agricultural production, even in technologically advanced societies, continues to be vulnerable to regionalized environmental (primarily climatic) abnormalities. However, in an age of increased global interdependence – not least, for the majority of states, with respect to basic food supplies – the impact of these events is quickly transmitted to international markets. Meanwhile, modification of the natural environment in the course of economic growth has reached such a scale that global consequences may flow from relatively localized (i.e. subcontinental) human interventions.

Chisholm (1982, p.177) argues that contemporary concern about climatic change and its implications can be traced to three separate sets of events: 'the accumulation of atmospheric carbon dioxide; the large number of extreme climatic events since 1960; and environmental degradation around the margins of deserts.' Although natural (exogenous) trends are involved in each of these phenomena, human activity is playing a major rôle, either in *causing* change or in increasing the *vulnerability* of particular populations to it (Blaikie 1985). Some of the pressures contributing to desertification were examined in Chapter 10, and comparable interdependencies between social and environmental systems exist in other ecological zones (Brookfield 1982). In general, 'ongoing underdevelopment is placing marginal people in marginal lands' (Susman, O'Keefe & Wisner 1983, p.280) so that, as the Chinese example suggests (see Ch. 2), the restoration of environmental productivity must be an early priority for any Third World regime committed to creating a more secure and equitably structured agrarian economy. The apparent increase in climatic and related 'hazards' which are disruptive and destructive of primary production is a global problem, however, from which capitalist and socialist societies suffer alike (Hewitt 1983a). It is noteworthy that the period of economic prosperity and stability which the USA enjoyed after 1945 coincided with 'unusually favourable [and] stable' climatic conditions on the Great Plains (Warrick 1983, p.67). Recent evidence of climatic warming and growing water deficits in the interior of the USA suggests that agricultural production faces a more difficult future (Brown & Postel 1987).

The increasing concentration of carbon dioxide (CO_2) in the atmosphere and depletion of the ozone layer are phenomena which have resulted primarily from human activity in the industrialized core nations, but which are likely to produce the most serious consequences within the economically and ecologically marginal economies of the periphery. The cumulative combustion of fossil fuels since the start of the Industrial Revolution is generally identified as the principal source of rising levels of atmospheric CO_2, although the spread of modern agricultural practices which reduce the organic carbon content of the soil may also be a substantial contributor (Chisholm 1982). Continuing deforestation especially in the Tropics, is reducing the capacity of the biosphere to remove atmospheric carbon dioxide. And although economic growth rates, and hence increases in energy consumption, have moderated since the 1950s and 1960s, the widespread public disfavour with which nuclear power plants have met in most Western nations has made public authorities correspondingly captive to fossil-fuel technologies for future energy supply.

Concern about rising levels of atmospheric CO_2 stems from its impact on the Earth's radiation balance. The so-called *greenhouse effect* is caused by the re-radiation back towards the Earth of some of the heat radiated upwards from the Earth's surface which is intercepted by the carbon dioxide in the atmosphere. No one can predict precisely how much, or how fast, average temperatures in the lower atmosphere will increase as a result, but it seems quite probable that major changes in the global climate will become evident during the course of the 21st century (US Government 1980). Indeed, some projections envisage significant developments by the year 2025. The consequences of rising temperatures and levels of atmospheric CO_2 are not all negative: Revelle (1982) suggests they are partly responsible for the improved agricultural yields which have been attained since the 1930s. However, they promise to become increasingly disruptive of existing economic activity, not only in the agricultural sector, but in urban–industrial communities also. Any significant warming of the atmosphere will likely be accompanied by increased seasonality (more climatic extremes) and uneven geographical incidence. Revelle notes, for example, that polar regions are expected to experience temperature increases up to three times as large as tropical latitudes, with uncertain consequences for sea-level change (as a result of melting ice caps) and the habitability of coastal regions. Changed distributions of precipitation and evapotranspiration could greatly reduce the flows of some river systems (the Colorado, Zambezi, and Hwang-Ho might be particularly vulnerable) and increase those of others (possibly the Niger, Volta, and Blue Nile).

Projections of climatic change in the foreseeable future abound with uncertainties. It is beyond dispute, however, that patterns of economic activity both in the global core and in the periphery are having a non-trivial impact on the environmental systems on which human life depends. Within every nation there are attitudes, and planned developments which they embody, which call for immediate reconsideration. For instance, large-scale water transfers from Arctic watersheds to water-deficit regions in the continental interior have long been proposed, both in North America and in the USSR. These would almost certainly have significant environmental impacts and yield very limited economic benefits. The Soviet plans appear finally to have been abandoned (Micklin 1986), but advocates of Canada–USA water transfers continue to dwell on what is technologically feasible, with little regard for environmental consequences (Scott *et al.* 1986). With respect to the disastrous shrinkage of the Aral Sea as a result of irrigation developments (Fig. 12.1), Micklin (1988, p.1172, draft version) observes: 'the spectrum and severity of accompanying environmental, ecological and socio-economic consequences have exceeded even the most pessimistic earlier predictions.'

The issues raised by military expenditure, the careless application of contemporary technologies, and the tolerance of environmental systems to large-scale human interventions come together particularly clearly in the western USA. Strategies which were proposed in the 1970s to deploy the MX missile system and to develop a commercial-scale synthetic-fuels industry were

> ... pursued independently by different federal agencies, with no regard for either the cumulative consequences of the two projects on the West, or for whether the pursuit of one project *may inadvertently jeopardize* the other (Rycroft & Monaghan 1982, p.69; italics added).

Together, the projects could have triggered a population influx of up to one million people in a region of limited urban infrastructure, have used up large amounts of land and scarce water, and demanded cement production equivalent to the entire capacity of the western USA.

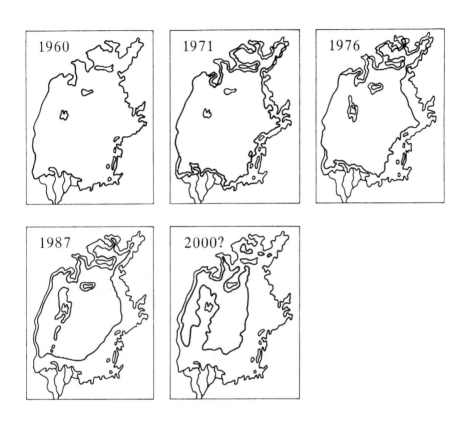

Year	Average Level (m)	Average Area (km^2)	Average Volume (km^3)	Average Salinity (g/l)
1960	53.41	68,000	1090	10
1971	51.05	60,200	925	n/a
1976	48.28	55,700	763	14
1987	40.50	41,000	374	27
2000	33.00	23,400	162	35

Figure 12.1   The changing profile of the Aral Sea.
Source: P. P. Micklin, Department of Geography, Western Michigan University, Kalamazoo.

Debt, employment and markets

For vast numbers of people, arms production on a scale which puts
survival itself in question, and environmental change that may
threaten the wellbeing of millions are remote threats. Making a
living is a more immediate and no less critical concern. The
experience of long-term unemployment or underemployment, or
of extremely unstable employment, is no longer confined to Third
World countries: it reappeared in Western core nations in the 1970s,
following three decades of relatively 'full' employment, and seems likely
to persist at varying levels of severity. Recent attempts to distinguish
temporary, and generally reversible, changes in the economy from the
evidence of permanent *structural* change, such as that associated with
a new Kondratieff cycle, have necessarily been somewhat conjectural
(Freeman *et al.* 1982). Some of the symptoms of economic malaise which
characterized the capitalist world-economy in the early 1980s disappeared
in the subsequent period of growth. But there are a number of reasons for
anticipating novel problems in the coming decades.

Some perceptive and not notably alarmist commentators have gone
so far as to suggest that current trends will raise fundamental questions
about the nature of work in advanced capitalist societies and the means
of allocating a nation's aggregate income among its inhabitants (Mouly
1979, Pahl 1980, Jones 1982). The widespread application of a new
generation of technologies could change the character of human work
tasks as radically as the 'first' Industrial Revolution. Then, the harnessing
of inanimate energy to the production process vastly extended the
ability of men and women to shape their material environment. The
modern revolution, based on microelectronics and the whole range
of 'intelligent' machines, vastly extends people's ability to exploit their
mental capacities and to order their informational environment. Simultaneously, in the form of *robotics*, it renders direct human involvement
in the processes of material transformation increasingly unnecessary.

Whether the overall impact of these technologies will increase
employment opportunities or decimate them is keenly and anxiously
debated. Trends in industrial output and manufacturing employment in
the core capitalist nations began to diverge in the mid-1960s (Fig.8.3).
Investment in production facilities was becoming increasingly governed
by competitive pressures to replace labour with capital equipment
as a means of reducing unit costs (see p.161), often augmented by
government policies which (perversely) tended to tax payrolls but to
subsidize investment. The potential undoubtedly exists for a further
drastic reduction in the core's manufacturing workforce, and in many

service occupations involving goods or routine information handling. The 'new international division of labour' (see Ch. 7) accentuates the problem, threatening core employment in relatively low-skill assembly operations, in the clothing or electronics industries for example. Such low-paying and labour-intensive operations are prime candidates either for relocation to the Third World or else for complete automation.

The prospect that capitalist economies would increase their output of goods and services while requiring the active participation of a diminishing proportion of their population has long appealed to many people (not least Marx). He envisaged a future age of leisure and freedom from tedious necessity. But it is already evident that within the political economy of contemporary capitalism the benefits accrue very unevenly. Those with relatively secure and remunerative 'good' jobs in growing, information-based or technology-intensive industries do well, whereas a greater number find that their employment is insecure, poorly paid, and without scope for advancement. Unemployment falls most heavily on the economically and politically powerless; young people, older workers, most women, migrant workers and, increasingly, those with below-average or unmarketable skills (OECD 1983). For those who face it, 'prolonged unemployment is ... a profoundly corrosive experience, undermining personality and atrophying work capabilities' (Harrison 1976, p.347). The poverty of permanent low-wage and/or intermittent employment produces similar stresses.

The severity of prolonged unemployment in the 1980s was much greater in Europe than in the USA, and there is no consensus as to how these differences are to be explained. A whole range of economic, political, and cultural variables (such as budgetary policy, rates of capital investment, forms of social security and attitudes to work and risk) seem to be involved. Yet at a time when occupational upheaval and structural economic change make social provision for vulnerable workers more necessary than for many decades, government response is limited: the State is either too much in debt to act effectively or, as in Britain under the Thatcher government, is ideologically disinclined to do so. Not surprisingly, there are many who argue that if government spending has to be limited, a reallocation of military budgets towards investment in people – through education, skill-training, and fostering the capacity for creative use of 'free' time – would arguably do more to reduce sources of conflict within and between nations than their current disposition.

Industrialized state-socialist societies are not immune from such problems. The 'full' employment which (until recently) officially existed in Soviet-style economies, in the sense that nearly everyone in the labour force 'had a job', was achieved at the cost of institutionalized inefficiency and underemployment (Adam 1982). Given the difficulties of improving labour utilization, the current slowdown in labour force *growth* within the USSR has major implications for that country's economic perform- ance during the rest of this century (Feshbach 1983). The question of how fast the USSR can afford to invest in the technology of the microelectronic revolution is related to a more fundamental one: Mr Gorbachev's *perestroika* notwithstanding, how actively will its controlling bureaucrats promote the diffusion of computer and telecommunications systems which encourage decentralized decision-making? In the East, as in the West, the opportunity costs of military spending in terms of social welfare have become increasingly apparent.

The global context of the core's employment problems is that the periphery faces an incomparably greater challenge of job-creation in the coming decades. Despite the recent declining trend in rates of population growth, it is clear that there will be a massive increase in the labour force of the Third World during the remainder of this century and beyond. Between 1975 and 2000 the low-income countries of Asia, for instance, will have gained an extra 250 mil- lion workers, an increment equal to their entire labour force in 1950 (Fig. 12.2; cf. Fig. 3.10). However, there are major obstacles to sustained expansion of formal sector employment in most Third World countries, and limits to the capacity of the informal sector continually to absorb extra workers (see p.255). Renshaw (1981) esti- mated that, given its annual rates of labour force increase (2.4%) and labour productivity increases (2.5%), 'the South' would require national income to grow by 4.9% annually just to keep unemployment from rising. Only a handful of Third World states attained such growth in the 1980s: the majority experienced a rapidly deteriorating economic environment, associated with the 'debt crisis' (George 1988).

After 1973, the major Western banks, as part of their recycling of OPEC 'petrodollars', assumed a direct rôle in financing Third World development. (Previously, most lending had come from multilateral agencies, usually on concessional terms.) During the 1970s, the real interest rate on this commercial debt was very low, but the situation changed rapidly after 1980. Interest rates increased dramatically, largely as a result of US action to cover *its* growing deficit, and recession in the industrialized Western nations cut their demand for commodity imports from Third World nations. The resulting fall in prices (see p.56), made it extremely difficult for most LDCs to service their debts. The 1979 oil price rise compounded the financial difficulties for most states, but so too did their pattern of spending. Enormous sums were sunk into grossly inappropriate (and usually unproductive) capital projects, equally unproductive armaments, conspicuous consumption by the corrupt élites of 'soft states' and, to top it all, 'capital flight' (the transfer of funds into private foreign bank accounts) by the same élites.

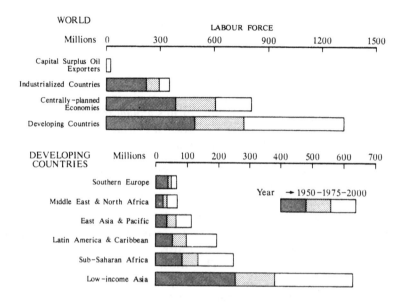

Figure 12.2   Labour force estimates and projections, 1950–2000.
Source: after World Bank 1979. *World development report 1979*. New York: Oxford University Press; Figure 7.

The efforts of the commercial banks to ensure that debtor nations do what is necessary to meet their interest charges have been coordinated by the International Monetary Fund (IMF). Its stock 'cure' is to require drastic reductions in State expenditure (although military spending is exempt!) and an all-out effort to expand exports. The typical effects of IMF 'adjustment' measures in the 1980s have been strong downward pressure on wages, drastic reductions in 'human capital' expenditures (health, education, etc.), the squeezing out of investment in non-export-oriented industry, formal-sector job losses, and a severe shrinkage of consumption. The 'benefits', of increased exports to meet the country's international financial obligations, have been considerably undermined by the imposition of the same 'remedy' on all debtors. Commodity prices have been depressed by excessive global production, and neomercantilism in the core has greatly hampered LDC efforts to expand manufactured exports (George 1988). The debt crisis thus compounds the 'employment crisis' facing most Third World nations.

The rationality of global armaments spending again asserts itself. The World Employment Conference of 1976 concluded that the only strategy that could hope to meet the Third World's requirements of job-creation and sustained economic growth was one which gave priority to eradicating poverty by investing in 'basic needs'; the social and economic infrastructures related to food supply, water supply, education, health care, and basic housing and transportation provision (Jolly 1976). As Barbara Ward (1976, p.270) observed:

> If we take the World Bank's [cost] estimate of [meeting]basic needs, we reach the remarkable conclusion that the entire proposed spending on the works of peace for an entire decade would amount to no more than half the world's *annual* bill for weapons. A yearly five percent transfer from arms spending to development would fund the entire World Bank program.

Whether it is desirable or not, the fortunes of most Third World states remain tied to the economic performance of the industrialized nations, and to the resource allocation decisions made by their governments and business corporations (Leontief & Sohn 1982). It is impossible to prescribe a development strategy for any underdeveloped country which is risk-free and unambiguously conducive to the greater welfare of its poorest citizens. The most effective strategy appears to combine an openness to the trading opportunities of the global economic system (not an *unregulated* exposure to its pressures), with positive measures to encourage widespread domestic participation in efficient productive activity, notably in agriculture. The rising prosperity of the leading NICs, and of many LDCs which aspire to emulate their industrialization, is critically dependent on the willingness of governments of Western core nations to respond to their own unemployment problems in ways which allow the volume of imports from the Third World to continue growing – no mean challenge!

In the absence of a new hegemony, the late 20th-century world-economy appears to be forming into a group of regional trading blocs. The EEC is becoming an increasingly integrated market, Japan shows signs of creating more formalized trading links within the Pacific region, and the Canadian and Mexican economies have grown more interdependent with that of the USA. It is a matter of global concern that these large regional markets should refrain from attempting to solve their internal problems by adopting protectionist measures against external parties. Perhaps recent experience, when North America and Western Europe lost an estimated 2–3 million jobs between 1981 and 1983 because of the reduced buying power of indebted Third World nations (George 1988), will foster the recognition that self-interest alone would point to widening the market for LDC products, not least those that compete with current First World output. Capital exports from the core to the periphery will still be necessary, but the outlook for official development assistance from the core 'falls far short of the needs of developing countries, especially of the low-income countries, if world poverty is to be seriously tackled' (World Bank 1983, p.34).

Legitimizing social change in a dynamic global economy

The contemporary fragmentation of political and economic influence, together with the decisively greater *trans*nationality of the modern TNC, underline the fact that the world-economy has become riddled with interdependencies and yet at the same time become fundamentally 'unmanageable' (Drucker 1986). No single institution, whether it be a superpower, a leading corporation, or a supranational agency such as the United Nations, possesses the combination of economic and strategic power and political legitimacy to coordinate the workings of the global economic system, and to resolve the inevitable conflicts of interest which arise between one state and another, and between states and transnational corporations.

The benefits of postwar Keynesian macro-economic management and the creation of the welfare state allowed Western core governments to claim credit for successfully eliminating the adverse consequences of capitalist economic development. Unemployment and abject poverty were shown no longer to be inevitable, and a growing national income could be guaranteed. This allowed everyone to share, albeit unequally, a rising standard of living, and made it possible for specific redistributive measures (including 'regional policies') to be undertaken without any immediate harm to capital accumulation. Since the early 1970s, however, these achievements have proved somewhat less than permanent. A more intensely competitive international market and heightened domestic expectations have squeezed the capacity of national governments to protect their citizens from the adverse consequences of changes in the global economy. The legitimacy of the social order has correspondingly been opened to more serious questioning (Habermas 1976, Hirsch 1976a). In a comparable way, economic stagnation and its frustrating daily consequences have undermined the popular reputation of state-socialist regimes. The majority of Third World governments, for reasons outlined in Chapter 5, equally face legitimation problems, heightened by their dramatically greater economic difficulties and frequently their non-democratic origins.

The dilemmas facing contemporary societies ultimately reflect particular world-views which define the nature of men and women, and the basis of their social and environmental relationships. Both the major contemporary world-views are flawed (Berger 1974). Capitalism has proved too willing to sacrifice human dignity in the interests of an elusive and narrowly defined economic efficiency. State-socialism ('genuine' communism remains utopian) has proved too willing to sacrifice human freedom and creativity in the interests of an imperfectly equitable political economy. Neither provides a satisfactory basis for resolving what Hirsch (1976a, p.190) defines as 'the prime economic problem now facing the economically advanced societies – a structural need to pull back from the bounds of economic self-advancement.'

The balance between individual initiative and mutual accountability in economic and social life remains elusive (Harvey 1985b, ch. 5). Despite the pressing need for global action to respond constructively to the issues raised in this chapter – the threats to social wellbeing posed by environmental change and degradation, militarism, and widespread poverty and unemployment – contemporary societies appear to be increasingly prone to becoming paralysed by pluralism or else succumbing to an ideology of the powerful. In neither case are the issues which need to be addressed, if economic development is to proceed in environmentally sustainable and humanly satisfying configurations, adequately resolved. Writers from many different backgrounds, including Hirsch, the social democrat, and Habermas, the (Marxian) critical theorist agree that the challenges facing late-20th-century societies call for reasoned ethical choices even more than they require imaginative technical solutions (Ellis & Kumar 1983).

The relationship between individuals and social structures and institutions is a reflexive one: we are shaped by our cultural environment but have the capacity, in turn, to shape it (Gregory 1981). The evolution of the capitalist world-economy attests to the power and durability of its underlying cultural appeal (see Ch. 1). But at the same time, '[modern] society is in turmoil because the only legitimacy it has is social justice' (Hirsch 1976a, p.190). The most hopeful future for the global economic system requires actions that attend to the twin goals of efficiency and equity. These are not necessarily mutually exclusive, although the focus will shift – uneasily at times no doubt – between the two. A similarly untidy combination of attention to global and to national (regional, local) priorities will be necessary. Do our societies have the cultural resources to accept these challenges? An understanding of the evolution and present condition of the global economic system, such as this book has attempted to provide, is a useful beginning. In an interdependent world, localized misconceptions (or plain ignorance) can be costly. The environmental security and economic prosperity of the peoples of the world is something about which geographers have informed and important things to say. Let us work to apply them.

Bibliography

Abouchar, A. 1979. *Economic evaluation of Soviet socialism.* New York: Pergamon Press.

Abrams, C. 1977. Squatting and squatters. In *Third World Urbanization*, J. Abu-Lughod & R. Hay (eds), 293–9. Chicago: Maaroufa Press.

Abrams, P. 1978. Introduction. In *Towns in societies: essays in economic history and historical sociology*, P. Abrams & E.A. Wrigley (eds), 1–7. Cambridge: Cambridge University Press.

Adam, J., (ed.) 1982. *Employment policies in the Soviet Union and Eastern Europe.* London: Macmillan.

Al-Chalabi, F. J. 1980. *OPEC and the international oil industry: a changing structure.* Oxford: Oxford University Press, for the Organization of Arab Petroleum Exporting Countries.

Armengaud, A. 1973. Population in Europe 1700–1914. In *The Fontana economic history of Europe*, vol. 3: *The Industrial Revolution 1700–1914*, C. M. Cipolla (ed.), 22–76. Glasgow: Fontana.

Armstrong, H. & J. Taylor 1978. *Regional economic policy and its analysis.* Oxford: Philip Allan.

Auty, R. M. 1984. The product life-cycle and the location of the global petrochemical industry after the second oil shock. *Economic Geography* 60; 325–38.

Auty, R. M. 1985. Export base theory, staple flexibility and tropical regional development. *Singapore Journal of Tropical Geography* 6, 13–22.

Ayeni, B. 1981. Lagos. In *Problems and planning in Third World cities*, M. Pacione (ed.). 127–55 London: Croom Helm.

Bagchi, A. K. 1976. De-industrialization in India in the nineteenth century: some theoretical implications. *Journal of Development Studies* 12, 135–64.

Bairoch, P. 1973. Agriculture and the Industrial Revolution 1700–1914. In *The Fonatana economic history of Europe*, Vol. 3: *The Industrial Revolution 1700–1914*, C. M. Cipolla (ed.), 452–506. Glasgow: Fontana.

Bairoch, P. 1975. *The economic development of the Third World since 1900.* London: Methuen.

Baran, P. A. 1957. *The political economy of growth.* New York: Monthly Review Press.

Barnet, R. J. & R. E. Muller 1974. *Global reach: the power of the multinational corporations.* New York: Simon and Schuster.

Barr, B. M. 1974. The changing impact of industrial management and decision-making on the locational behaviour of the Soviet firm. In *Spatial perspectives on industrial organization and decision-making*, F. E. I. Hamilton (ed.), 411–46. Chichester: Wiley.

Barraclough, G. (ed.) 1978. *The Times atlas of world history*. London: Times Books.

Bartels, C. P. A. and J. J. van Duijn 1982. Regional economic policy in a changed labour market. *Papers of the Regional Science Association* 49, 97–111.

Bater, J. H. 1977. Soviet town planning: theory and practice in the 1970s. *Progress in Human Geography* 1, 177–207.

Bater, J. H. 1979. The legacy of autocracy: environmental quality in St. Petersburg. *The socialist city: spatial structure and urban policy*, R. A. French & F. E. I. Hamilton (eds), 23–47. Chichester: Wiley.

Bates, R. H. 1979. The commercialization of agriculture and the rise of rural political protest in Black Africa. In *Food, politics and agricultural development*, R. F. Hopkins, D. J. Puchala & R. B. Talbot (eds) 227–59. Denver, Col.: Westview.

Beaver, S. H. 1967. Ships and shipping: the geographical consequences of technological progress. *Geography* 52, 133–56.

Berger, P. L. 1974. *Pyramids of sacrifice: political ethics and social change*. New York: Basic Books.

Bergmann, T. 1975. *Farm policies in socialist countries*. Farnborough, Hants.: Saxon House.

Bergson, A. & H. S. Levine (eds) 1983. *The Soviet economy: toward the year 2000*. London: Allen & Unwin.

Berry, B. J. L. 1980. Inner city futures: An American dilemma revisited. *Institute of British Geographers, Transactions NS* 5, 1–28.

Berry, B. J. L., E. C. Conklin & D. M. Ray 1976. *The geography of economic systems*. Englewood Cliffs, NJ: Prentice-Hall.

Beyers, W. B. 1979. Contemporary trends in the regional economic development of the United States. *Professional Geographer* 31, 34–44.

Bienefeld, M. 1981. Dependency and the newly industrialising countries (NICs): towards a reappraisal. In *Dependency theory: a critical reassessment*, D. Seers (ed.), 79–96. London: Frances Pinter.

Birch, D. L. 1979. *The job generation process*. Cambridge, Mass.: MIT Press.

Blackaby, F. (ed.) 1978. *De-industrialisation*. London: Heinemann, for the National Institute of Economic and Social Research.

Blaikie, P. 1985. *The political economy of soil erosion in developing countries*. London: Longman.

Blaikie, P. 1986. Natural resource use in developing countries. In *A world in crisis? Geographical perspectives*, R. J. Johnston & P. J. Taylor, (eds), 107–26. Oxford: Blackwell.

Bluestone, B. & B. Harrison 1982. *The deindustrialisation of America*. New York: Basic Books.

Bond, A. R. 1987. Spatial dimensions of Gorbachev's economic strategy. *Soviet Geography* 28, 490–523.

Bond, A R. 1988. Moscow under restructuring: introduction to the special issue. *Soviet Geography* 29, 1–15.

Borchert, J. R. 1967. American metropolitan evolution. *Geographical Review* 57, 301–32.

Borchert, J. R. 1978. Major control points in American economic geography. *Annals, Association of American Geographers* 68, 214–32.

Boserup, E. 1965. *The conditions of agricultural growth: the economics of agrarian change under population pressure*. Chicago: Aldine.

Boulding, K. E. 1970. *Beyond economics; essays on society, religion and ethics*. Ann Arbor: University of Michigan Press.

Bowers, J. K. & P. Cheshire 1983. *Agriculture, the countryside and land use: an economic critique*. London: Methuen.

Bradford, C. I. 1982. Rise of the NICs as exporters on a global scale. In *The newly industrializing countries: trade and adjustment*, L. Turner & N. McMullen (eds), 7–24. London: Allen & Unwin.

Bradshaw, M. J. 1988. Soviet Asian–Pacific trade and the regional development of the Soviet Far East. *Soviet Geography* 29, 367–93.

Brandt, W. 1980. *North–South: a program for survival – report of the Independent Commission on International Development Issues*. Cambridge, Mass.: MIT Press.

Breheny, M., P. Cheshire & R. Langridge 1985. The anatomy of job creation? Industrial change in Britain's M4 corridor. In *Silicon landscapes*, P. Hall & A. Markusen (eds), 118–33. London: Allen & Unwin.

Brewis, T. N. 1969. *Regional economic policies in Canada*. Toronto: Macmillan.

Briggs, A. 1968. *Victorian cities*. London: Penguin.

Britton, J. N. H. 1976. The influence of corporate organization and ownership on the linkages of industrial plants: a Canadian enquiry. *Economic Geography* 52, 127–41.

Britton, J. N. H. & J. M. Gilmour 1978. *The weakest link: a technological perspective on Canadian industrial underdevelopment*. Background Study 43. Ottawa: Science Council of Canada.

Brookfield, H. 1975. *Interdependent development*. London: Methuen.

Brookfield, H. 1982. On man and ecosystems. *International Social Science Journal* 34, 375–93.

Browett, J. 1985. The newly industrializing countries and radical theories of development. *World Development* 13, 789–803.

Brown, L. R. 1985. A false sense of security. In *State of the world – 1985*, L. R. Brown et al., 3–22. New York: Norton.

Brown, L. R. & J. Jacobson 1987 Assessing the future of urbanisation. In *State of the world – 1987*, L. R. Brown et al., 38–56. New York: Norton.

Brown, L. R. & S. Postel 1987. Threshold of change. In *State of the world – 1987*, L. R. Brown et al., 3–19. New York: Norton.

Browning, C. E. 1981. Federal outlays and regional development. In *Federalism and regional development: case studies on the experience in the United States and the Federal Republic of Germany*, G. W. Hoffman (ed.), 125–53. Austin: University of Texas Press.

Browning, C. E. & W. Gesler 1979. The Sun Belt–Snow Belt: a case of sloppy regionalizing. *Professional Geographer* 31, 66–74.

Bruchey, S. 1965. *The roots of American economic growth, 1607–1861: an essay in social causation*. New York: Harper & Row.

Buchanan, K. 1970. *The transformation of the Chinese earth*. London: Bell.

Burgess, J. A. 1982. Selling places: Environmental images for the executive. *Regional Studies* 16, 1–17.

Burghardt, A. F. 1971. A hypothesis about gateway cities. *Annals, Association of American Geographers* 61, 269–85.

Buswell, R. J. & E. W. Lewis 1970. The geographical distribution of industrial research activity in the UK. *Regional Studies* 4, 297–306.

Buttel, F. H., M. Kenney & J. Koppenburg, Jr 1985. From Green Revolution to biorevolution: some observations on the changing technological bases of economic transformation in the Third World. *Economic Development and Cultural Change* 34, 31–55.

Calcagno, A. E. & J. Knakal 1981. Transnational companies and direct private investment in developing countries. In *East–West–South: economic interactions between three worlds*, C. T. Saunders (ed.), 103–26. London: Macmillan.

Chenery, H. B. 1980. Poverty and progress – choices for the developing world. *Finance and Development* **17**(2), 12–16.

Chisholm, M. 1976. Regional policies in an era of slow population growth and higher unemployment. *Regional Studies* **10**, 201–13.

Chisholm, M. 1979. *Rural settlement and land use: an essay in location* 3rd ed. London: Hutchinson.

Chisholm, M. 1982. *Modern world development: a geographical perspective.* London: Hutchinson.

Chisholm, M. 1987. Regional development: the Reagan–Thatcher legacy. *Government and Policy* **5**, 197–218.

Chung, R. 1970. Space–time diffusion of the transition model: the twentieth century patterns. In *Population geography: a reader*, G. J. Demko *et al.* (eds), 220–39. New York: McGraw–Hill.

Clark, G. & M. Dear 1981. The state in capitalism and the capitalist state. In *Urbanization and urban planning in capitalist society*, M. Dear & A. J. Scott (eds), 45–61. London: Methuen.

Clem, R. S. 1980. Regional patterns of population change in the Soviet Union 1959–1979. *Geographical Review* **70**, 137–56.

Collins, L. 1966. *Industrial migration and relocation: a study of European branch plants with special reference to metropolitan Toronto.* MA thesis, Department of Geography, University of Toronto (unpubl.).

Colman, D. & F. Nixson 1978. *Economics of change in less developed countries.* Oxford: Philip Allan.

Conference Board 1981. *Major forces in the world economy: concerns for international business.* Research report no. 807. New York: The Conference Board Inc.

Cook, E. 1976. *Man, energy, society.* San Francisco: W. H. Freeman.

Cooke, P. 1988. Flexible integration, scope economies, and strategic alliances: social and spatial mediations. *Society and Space* **6**, 281–300.

Corbridge, S. 1986. *Capitalist world development: a critique of radical development geography.* Totowa, NJ: Rowman and Allanheld.

Courtenay, P. P. 1980. *Plantation agriculture*, 2nd edn. London: Bell and Hyman.

Daniels, P. W. 1977. Office location in the British conurbations: trends and strategies. *Urban Studies* **14**, 261–74.

Darby, H. C. 1936. *An historical geography of England before A.D. 1800.* Cambridge: Cambridge University Press.

Datoo, B. A. 1978. Toward a reformulation of Boserup's theory of agricultural change. *Economic Geography* **54**, 135–44.

Daunton, M. J. 1978. Towns and economic growth in eighteenth-century England. In *Towns in Societies: Essays in Economic History and Historical Sociology*, P. Abrams & E.A. Wrigley (eds), 245–77. Cambridge: Cambridge University Press.

David, P. A. 1975. *Technical choice, innovation and economic growth.* Cambridge: Cambridge University Press.

Davidson, B. 1978. *Africa in modern history: the search for a new society.* London: Allen Lane.

Denecke, D. 1976. Innovation and diffusion of the potato in Central Europe in the seventeenth and eighteenth centuries. In *Fields, farms and settlement in Europe*, R. H. Buchanan, R. A. Butlin & D. McCourt (eds), 60–96. Belfast: Ulster Folk and Transport Museum.

Denison, E. F. 1967. *Why growth rates differ: postwar experience in nine Western countries*. Washington: Brookings Institution.

Dennis, R. 1978. The decline of manufacturing employment in Greater London: 1966–74. *Urban Studies* 15, 63–73.

Denoon, D. 1983. *Settler Capitalism: the dynamics of dependent development in the Southern Hemisphere*. Oxford: Clarendon Press.

De Vries, J. 1974. *The Dutch rural economy in the golden age, 1500–1700*. New Haven, Conn.: Yale University Press.

Dicken, P. 1986. *Global shift: industrial change in a turbulent world*. London: Harper & Row.

Dicken, P. 1987. Japanese penetration of the European automobile industry: the arrival of Nissan in the United Kingdom. *Tijdschrift voor Economische en Sociale Geografie* 78, 59–72.

Dicken, P. 1988. The changing geography of Japanese foreign direct investment in manufacturing industry: a global perspective. *Environment and Planning A* 20, 633–53.

Dicken, P. & P. E. Lloyd 1976. Geographical perspectives on United States investment in the United Kingdom. *Environment and Planning A* 8, 685–705.

Dicken, P. & P. E. Lloyd 1981. *Modern Western society: a geographical perspective on work, home and well-being*. London: Harper & Row.

Dos Santos, T. 1973. The crisis of development theory and the problems of dependence in Latin America. In *Underdevelopment and development*, H. Bernstein (ed.), 57–80. London: Penguin.

Douglas, I. 1983. *The urban environment*. London: Edward Arnold.

Drakakis-Smith, D. 1981. *Urbanisation, housing and the development process*. London: Croom Helm.

Drucker, P. F. 1986. The changed world economy. *Foreign Affairs*, 20, 768–91.

Dunford, M. & D. Perrons 1983 *The arena of capital*. London: Macmillan.

Dunning, J. H. & J. M. Stopford 1983. *Multinationals: company performance and global trends*. London: Macmillan.

Economic Council of Canada 1977. *Living Together: a study of regional disparities*. Ottawa: Supply and Services Canada.

Edwards, G. A. 1977. Historical perspective on acquisition trends. *Foreign Investment Review* (Ottawa) 1(1), 12–13.

Ellis, A. & K. Kumar (eds) 1983. *Dilemmas of liberal democracies: studies in Fred Hirsch's 'social limits to growth'*. London: Tavistock.

Ellis, F. 1982. Prices and the transformation of peasant agriculture: the Tanzanian case. *IDS Bulletin*, Sussex 13(4), 66–72.

Ellul, J. 1978. *The betrayal of the West*. New York: Seabury Press.

Estall, R. 1982. Planning in Appalachia: an examination of the Appalachian Regional Development Programme and its implications for the future of the American Regional Planning Commissions. *Institute of British Geographers, Transactions NS* 7, 35–58.

Eversley, D. 1980. Employment for the inner city. *Institute of British Geographers, Transactions NS* 5, 141–50.

Eyre, S. R. 1978. *The real wealth of nations*. London: Edward Arnold.

Feder, E. 1980. The new agrarian and agricultural change trends in Latin America. In *Environment, society, and rural change in Latin America*, D. A.

Preston (ed.), 211–31. New York: Wiley.

Ferraro, V., E. Doherty & B. Cassani 1982. The resource costs of global poverty: an analytical review of the literature. *International Journal of Social Economics* 9(5), 3–35.

Feshbach, M. 1983. Population and labour force. In *The Soviet economy: toward the year 2000*, A. Bergson & H. S. Levine (eds), 79–111. London: Allen & Unwin.

Field, N. C. 1968. Environmental quality and land productivity: a comparison of the agricultural land base of the USSR and North America. *Canadian Geographer* 12, 1–14.

Flinn, M. W. 1966. *The origins of the Industrial Revolution*. London: Longman.

Flinn, M. W. 1967. Social theory and the Industrial Revolution. In *Social theory and economic change*, T. Burns & S. B. Saul (eds), 9–34. London: Tavistock.

Flinn, M. W. 1978. Technical change as an escape from resource scarcity: England in the seventeenth and eighteenth centuries. In *Natural resources in European history*, A. Maczak & W. N. Parker (eds), 139–59. Washington: Resources for the Future.

Flynn, N. & A. P. Taylor 1986. Inside the rust belt: an analysis of the decline of the West Midlands economy. 1: International and national economic conditions. *Environment and Planning A* 18, 865–900.

Frank, A. G. 1966. The development of underdevelopment. *Monthly Review* 18 (4), 17–31.

Frank, A. G. 1975. Development and underdevelopment in the New World: Smith and Marx vs. the Weberians. *Theory and Society* 2, 431–67.

Frank, A. G. 1979. *Dependent accumulation and underdevelopment*. New York: Monthly Review Press.

Franko, L. G. & S. Stephenson 1982. The micro picture: corporate and sectoral developments. In *The newly industrializing countries: trade and adjustment*, L. Turner & N. McMullen (eds), 193–219. London: Allen & Unwin.

Freeman, C. 1987. Technical innovation, long cycles and regional policy. In *Technical change and industrial policy*, K. Chapman & G. Humphrys (eds), 10–25. Oxford: Basil Blackwell.

Freeman, C., J. Clark & L. Soete 1982. *Unemployment and technical innovation: a study of long waves and economic development*. London: Frances Pinter.

French, R. A. & F. E. I. Hamilton 1979. Is there a socialist city? In *The socialist city: spatial structure and urban policy*, R. A. French & F. E. I. Hamilton (eds), 1–21. Chichester: Wiley.

Friedland, R. 1981. Central city fiscal strains: the public costs of private growth. *International Journal of Urban and Regional Research* 5, 356–76.

Friedmann, J. & C. Weaver 1979. *Territory and function: the evolution of regional planning*. London: Edward Arnold.

Friedmann, J. & G. Wolff 1982. World city formation: an agenda for research and action. *International Journal of Urban and Regional Research* 6, 309–43.

Friedmann, J. & R. Wulff 1979. *The urban transition: comparative studies of newly industrializing societies*. London: Edward Arnold.

Fröbel, F., J. Heinrichs & O. Kreye 1980. *The new international division of labour: structural unemployment in industrialised countries and industrialisation in developing countries*. Cambridge: Cambridge University Press.

Frost, M. & N. Spence 1981. Policy responses to urban and regional economic change in Britain. *Geographical Journal* 147, 321–49.

Fuchs, R. J. & G. J. Demko 1979. Geographic inequality under socialism. *Annals, Association of American Geographers* 69, 304–18.

Galbraith, J. K. 1973. *Economics and the public purpose.* Boston: Houghton Mifflin.

Garreau, G. 1977. *L'agrobusiness.* Paris: Calmann–Levy.

Geertz, C. 1963. *Agricultural involution: the process of ecological change in Indonesia.* Berkeley: University of California Press.

George, R. 1981. Cape Breton Development Corporation. In *Public corporations and public policy in Canada,* A. Tupper & G. B. Doern (eds), 365–88. Montreal: Institute for Research on Public Policy.

George, S. 1988. *A fate worse than debt.* London: Penguin.

Gerry, C. 1979. Small-scale manufacturing and repairs in Dakar: a survey of market relations within the urban economy. In *Casual work and poverty in Third World cities,* R. Bromley & C. Gerry (eds), 229–50. Chichester: Wiley.

Gershuny, J. 1978. *After industrial society: the emerging self-service economy.* London: Methuen.

Giddens, A. 1981. *A contemporary critique of historical materialism,* Vol. 1: *Power, property and the state.* London: Macmillan.

Gilbert, A. G. & D. E. Goodman 1976. Regional income disparities and economic development: a critique. In *Development planning and spatial structure,* A. G. Gilbert (ed), 113–38. Chichester: Wiley.

Goodzwaard, B. 1979. *Capitalism and progress: a diagnosis of Western society.* Grand Rapids, Michigan: Eerdmans.

Gordon, D. M. 1978. Capitalist development and the history of American cities. In *Marxism and the metropolis: new perspectives in urban political economy,* W. K. Tabb & L. Sawers (eds), 25–63. New York: Oxford University Press.

Gore, C. 1984. *Regions in question: space, development theory and regional policy.* London: Methuen.

Gorlov, V. N. & V. L. Baburin 1985. The new Soviet industrial corporations and research-based industrial corporations as economic–geographic study objects. *Soviet Geography* 26, 177–83.

Gottman, J. 1966. Why the skyscraper? *Geographical Review* 56, 190–212.

Gould, P. 1963. Man against his environment: a game-theoretic framework. *Annals, Association of American Geographers* 53, 290–7.

Goulet, D. 1973. *The cruel choice: a new concept in the theory of development.* New York: Atheneum.

Goulet, D. 1980. Development experts: the one-eyed giants. *World Development* 8, 481–9.

Graham, E. & I. Floering 1984. *The modern plantation in the Third World.* London: Croom Helm.

Grampp, W. D. 1970. On manufacturing and development. *Economic Development and Cultural Change* 18, 451–63.

Gregor, H. F. 1982. Large-scale farming as a cultural dilemma in U.S. rural development – the role of capital. *Geoforum* 13, 1–10.

Gregory, D. 1981. Human agency and human geography. *Transactions, Institute of British Geographers NS* 6, 1–18.

Griffin, K. 1970a. Reform and diversification in a coffee economy: the case of Guatemala. In *Unfashionable economics: essays in honour of Lord Balogh,* P. Streeten (ed.), 75–97. London: Weidenfeld and Nicolson.

Griffin, K. 1970b. Foreign capital, domestic savings and economic development. *Oxford Institute of Economics and Statistics*, Bulletin 32, 99–112.

Grigg, D. 1970. *The harsh lands: a study in agricultural development*. London: Macmillan.

Grigg, D. B. 1974. *The agricultural systems of the world: an evolutionary approach*. Cambridge: Cambridge University Press.

Grigg, D. 1979. Esther Boserup's theory of agrarian change: a critical review. *Progress in Human Geography* 3, 64–84.

Habermas, J. 1976. *Legitimation crisis*. London: Heinemann.

Hall, J. M. 1970. Industry grows where the grass is greener. *Area* 2 (3), 40–6.

Hall, P. 1980. *Great planning disasters*. London: Weidenfeld & Nicolson.

Hall, P. 1985. The geography of the fifth Kondratieff. In *Silicon landscapes*, P. Hall & A. Markusen (eds), 1–19. London: Allen & Unwin.

Hall, P. & A. Markusen 1985. *Silicon landscapes*. London: Allen & Unwin.

Hall, P., H. Gracey, R. Drewitt, & L. R. Thomas 1973. *The containment of urban England* (2 vols). London: Allen & Unwin.

Hamilton, F. E. I. 1971. Decision-making and industrial location in Eastern Europe. *Institute of British Geographers, Transactions* No. 52, 77–94.

Hamilton, F. E. I. 1974. Self-management: the Yugoslav case. In *Spatial perspectives on industrial organization and decision-making*, F. E. I. Hamilton (ed.), 449–59. Chichester: Wiley.

Hamilton, F. E. I. 1976. Multinational enterprise and the European Economic Community, *Tijdschrift voor Economische en Sociale Geografie* 67, 258–78.

Hare, F. K., R. W. Kates & A. Warren 1977. The making of deserts: climate, ecology, and society. *Economic Geography* 53, 332-46.

Harris, R. C. 1966. *The seigneurial system in early Canada: a geographical study*. Madison: University of Wisconsin Press.

Harris, R. C. & L. Guelke 1977. Land and society in early Canada and South Africa. *Journal of Historical Geography* 3, 135–53.

Harrison, B. 1982. The tendency toward instability and inequality underlying the "revival" of New England. *Papers of the Regional Science Association* 50, 41–65.

Harrison, R. 1976. The demoralising experience of prolonged unemployment. *Employment Gazette* (April), 339–48.

Harvey, D. 1974. Population, resources, and the ideology of science. *Economic Geography* 50, 256–77.

Harvey, D. 1982. *The limits to capital*. Oxford: Basil Blackwell.

Harvey, D. 1985a. *The urbanization of capital: studies in the history and theory of capitalist urbanization*. Baltimore: Johns Hopkins University Press.

Harvey, D. 1985b. *Consciousness and the urban experience: studies in the history and theory of capitalist urbanization*. Baltimore: Johns Hopkins University Press.

Hausner, V. A. 1987. Introduction: economic change and urban policy. In *Urban economic change: five city studies*, V. A. Hausner (ed.), 1–43. Oxford: Clarendon Press.

Hayter, R. 1982. Truncation, the international firm and regional policy. *Area* 14, 277–82.

Hayter, T. 1971. *Aid as imperialism*. London: Penguin.

Hebden, R. E. 1980. Trends in Soviet trade since 1960. *Geography* 65, 49–52.

Hecht, S. B. 1985. Environment, development and politics: capital accumulation and the livestock sector in Eastern Amazonia. *World Development* 13, 663–84.

Helleiner, G. K. 1982. International trade theory and Northern protectionism against Southern manufacturers. In *For good or evil: economic theory and North–South negotiations*, G. K. Helleiner (ed.), 47–61. Toronto: University of Toronto Press.

Herman, E. S. 1981. *Corporate control, corporate power*. New York: Cambridge University Press.

Hermassi, E. 1980. *The third world reassessed*. Berkeley: University of California Press.

Hewitt, K. 1983a. Interpreting the role of hazards in agriculture. In *Interpretations of calamity*, K. Hewitt (ed.), 123–39. Boston: Allen & Unwin.

Hewitt, K. 1983b. Place annihilation: area bombing and the fate of urban places. *Annals, Association of American Geographers* 73, 257–84.

Hirsch, F. 1976a. *Social limits to growth*. Cambridge, Mass.: Harvard University Press.

Hirsch, F. 1976b. Is there a new international economic order? *International Organisation* 30, 521–31.

Hirsch, F. 1978. The ideological underlay of inflation. In *The political economy of inflation*, F. Hirsch & J. H. Goldthorpe, (eds), 263–84. Cambridge, Mass.: Harvard University Press.

Hopkins, A. G. 1973. *An economic history of West Africa*. New York: Columbia University Press.

Hopkins, A. G. 1978. Imperial connections. In *The imperial impact: studies in the economic history of Africa and India*, C. Dewey & A. G. Hopkins (eds), 1–19. London: Athlone Press.

Horowitz, I. L. 1982. *Beyond empire and revolution: militarism in the Third World*. New York: Oxford University Press.

Howells, J. & A. E. Green 1986. Location, technology and industrial organisation in U.K. Services. *Progress in Planning* 26, 83–184.

Howells, J. R. L. 1984. The location of research and development: some observations and evidence from Britain. *Regional Studies* 18, 13–29.

Jensen, R. G. 1969. Regionalization and price zonation in Soviet agricultural planning. *Annals, Association of American Geographers* 59, 324–47.

Jolly, R. 1976. The world employment conference: the enthronement of basic needs. *Overseas Development Institute Review* (2), 31–44.

Jolly, R. 1978. Objectives and means for linking disarmament to development. In *Disarmament and world development*, R. Jolly (ed.), 105–12. Oxford: Pergamon Press.

Jones, B. 1982. *Sleepers, wake! Technology and the future of work*. Brighton: Wheatsheaf Books.

Kamarck, A. M. 1976. *The Tropics and economic development: a provocative inquiry into the poverty of nations*. Baltimore: Johns Hopkins University Press.

Kammen, M. 1970. *Empire and interest: the American colonies and the politics of mercantilism*. Philadelphia: Lippincott.

Karcz, J. F. 1979. *The economics of communist agriculture: selected papers*. Bloomington, Ind.: International Development Institute.

Keeble, D. 1976. *Industrial location and planning in the United Kingdom*. London: Methuen.

Keeble, D. E. 1980. Industrial decline, regional policy and the urban–rural manufacturing shift in the United Kingdom. *Environment and Planning A* 12, 945–62.

Keesing, D. B. 1979. *Trade policy for developing countries*. Staff Working Paper No. 353 Washington, DC. World Bank.

Kemp, T. 1969. *Industrialization in nineteenth-century Europe*. London: Longman.

Keynes, J. M. 1936. *The general theory of employment, interest and money*. London: Macmillan.

Khinchuk, K. 1987. Agricultural labour force in the Soviet Union. *Soviet Geography* 28, 90–115.

King, A. D. 1976. *Colonial urban development: culture, social power and environment*. London: Routledge & Kegan Paul.

Koropeckyj, I. S. 1967. The development of Soviet location theory before the Second World War. *Soviet Studies* 19, 1–28, 232–44.

Kutcher, G. P. & P. L. Scandizzo 1982. *The agricultural economy of northeast Brazil*. Baltimore: Johns Hopkins University Press.

Kuznets, S. 1955. Economic growth and income inequality. *American Economic Review* 45, 1–28.

Ladejinsky, W. 1976. Agricultural production and constraints. *World Development* 4, 1–10.

Lall, S. 1980. *The multinational corporation*. London: Macmillan.

Landes, D. S. 1969. *The unbound Prometheus: technological change, 1750 to the present*. Cambridge: Cambridge University Press.

Lappé, F. M. & J. Collins 1978. *Food first: beyond the myth of scarcity*. Boston: Houghton Mifflin.

Latham, A. J. H. 1978. *The international economy and the underdeveloped world 1865–1914*. London: Croom Helm.

Lemon, A. 1982. Migrant labour and frontier commuters: reorganizing South Africa's Black labour supply. In *Living under apartheid: aspects of urbanization and social change in South Africa*, D. M. Smith (ed.), 64–89. London: Allen & Unwin.

Leontief, W. & I. Sohn 1982. Economic growth. In *Population and the world economy in the 21st century*, J. Faaland (ed.), 96–127. Oxford: Basil Blackwell.

Lever, W. F. 1981. The inner-city employment problem in Great Britain since 1952: a shift-share approach. In *Industrial location and regional systems: spatial organization in the economic sector*, J. Rees, G. J. D. Hewings & H. A. Stafford (eds), 171–96. New York: J. F. Bergin.

Lewis, W. A. 1955. *The theory of economic growth*. London: Allen & Unwin.

Ley, D. 1983. *A social geography of the city*. New York: Harper & Row.

Lindert, P. H. 1986. *International economics*, 8th ed. Homewood, Ill.: Irwin.

Linge, G. J. R. & F. E. I. Hamilton 1981. International industrial systems. In *Spatial analysis, industry and the industrial environment: progress in research and applications*, Vol. II: *International industrial systems*, F. E. I. Hamilton & G. J. R. Linge (eds), 1–117. Chichester: Wiley.

Lipietz, A. 1986. New tendencies in the international division of labour: regimes of accumulation and modes of regulation. In *Production, work, territory*, A. J. Scott & M. Storper (eds), 16–40. Boston: Allen & Unwin.

Lipton, M. 1978. Inter-farm, inter-regional and farm–non-farm income distribution: the impact of the new cereal varieties. *World Development* 6, 319–37.

Lis, C. & H. Soly 1979. *Poverty and capitalism in pre-industrial Europe.* Hassocks, Sussex: Harvester Press.

Littwin, L. 1974. *Latin America: catholicism and class conflict.* Enrico, Cal.: Dickenson.

Loehr, W. & J. P. Powelson 1983. *Threat to development: pitfalls of the NIEO.* Boulder, Col.: Westview.

Lonsdale, R. E. & H. L. Seyler (1979). *Nonmetropolitan industrialization.* Washington, DC: Winston.

Luckham, R. 1978. Militarism and international dependence. In *Disarmament and world development*, R. Jolly (ed.), 35–56. Oxford: Pergamon Press.

Lydolph, P. E. 1979. *Geography of the U.S.S.R.: topical analysis.* Elkhart Lake, Wisc.: Misty Valley Publishing.

Mabogunje, A. L. 1973. Manufacturing and the geography of development in tropical Africa. *Economic Geography* 49, 1–20.

Mabogunje, A. L. 1980. *The development process: a spatial perspective.* London: Hutchinson.

Macfarlane, A. 1979. *The origins of English individualism: the family, property and social transition.* New York: Cambridge University Press.

Mackenzie, F. 1986. Local initiatives and national policy: gender and agricultural change in Murang'a District, Kenya. *Canadian Journal of African Studies* 20, 377–401.

Malecki, E. J. 1979. Locational trends in R and D by large US corporations, 1965–1977. *Economic Geography* 55, 309–23.

Malecki, E. J. 1980. Corporate organization of R and D and the location of technological activities. *Regional Studies* 14, 219–34.

Malecki, E. J. 1982. Federal R and D spending in the United States of America: some impacts on metropolitan economies. *Regional Studies* 16, 19–35.

Malthus, T. R. 1798/1914. *An essay on the principle of population.* (2 vols). London: Dent.

Mandel, E. 1968. *Marxist economic theory.* London: Merlin Press.

Mandel, E. 1980. *Long waves of capitalist development: the Marxist interpretation.* Cambridge: Cambridge University Press.

Markusen, A. R. 1986. Defence spending; a successful industry policy? *International Journal of Urban and Regional Research* 10, 105–22.

Martin, R. L. 1986. Thatcherism and Britain's industrial landscape. In *The geography of de-industrialisation*, R. Martin & B. Rowthorn (eds), Basingstoke: Macmillan.

Martin, R. L. & J. S. C. Hodge 1983. The reconstruction of British regional policy: 1. The crisis of conventional practice. *Environment and Planning C: Government and Policy* 1, 133–52.

Marx, K. & F. Engels 1848/1967. *The communist manifesto.* London: Penguin.

Mason, C. M. & R. T. Harrison 1985. The geography of small firms in the UK: towards a research agenda. *Progress in Human Geography* 9, 1–37.

Massey, D. 1979. In what sense a regional problem? *Regional Studies* 13, 233–43.

Massey, D. 1984. *Spatial divisions of labour: social structures and the geography of production.* London: Macmillan.

Massey, D. B. & R. A. Meegan 1978. Industrial restructuring versus the cities. *Urban Studies* 15, 273–88.

Massey, D. & R. Meegan 1982. *The anatomy of job loss: the how, why and where of employment decline.* London: Methuen.

Matley, I. M. 1966. The Marxist approach to the geographical environment. *Annals, Association of American Geographers* 56, 97–111.

Maxcy, G. 1981. *The multinational motor industry.* London: Croom Helm.

McCall, M. K. 1977. Political economy and rural transport: a reappraisal of transportation impacts. *Antipode* 9(1), 56–67.

McCann, L. D. (ed.) 1987. *Heartland and hinterland: a geography of Canada,* 2nd edn. Toronto: Prentice-Hall Canada.

McCarthy, D. M. P. 1982. *Colonial bureaucracy and creating underdevelopment: Tanganyika, 1919–1940.* Ames, Iowa: The Iowa State University Press.

McCloskey, D. N. 1975a. The persistence of English common fields. In *European peasants and their markets: essays in agrarian economic history,* W. N. Parker & E. L. Jones (eds), 73–119. Princeton, NJ: Princeton University Press.

McCloskey, D. N. 1975b. The economics of enclosure: a market analysis. In *European peasants and their markets; essays in agrarian economic history,* W. N. Parker & E. L. Jones (eds), 123–60. Princeton, NJ: Princeton University Press.

McConnell, J. E. 1980. Foreign direct investment in the United States. *Annals, Association of American Geographers* 70, 259–70.

McConnell, J. E. 1986. Geography of international trade. *Progress in Human Geography* 10, 471–83.

McCrone, G. 1969. *Regional policy in Britain.* London: Allen & Unwin.

McEnery, J. H. 1981. *Manufacturing two nations: the sociological trap created by the bias of British regional policy against service industry.* Research Monograph 36. London: Institute of Economic Affairs.

McGee, T. G. 1971. *The urbanization process in the Third World: explorations in search of a theory.* London: Bell.

McGee, T. G. 1977a. The persistence of the proto-proletariat: occupational structures and planning of the future of Third World cities. In *Third World urbanization,* J. Abu-Lughod & R. Hay, Jr. (eds), 257–70 Chicago: Maaroufa Press.

McGee, T. G. 1977b. Rural–urban mobility in South and Southeast Asia. Different formulations . . . different answers? In *Third World Urbanization,* J. Abu-Lughod & R. Hay, Jr. (eds), 196–212. Chicago: Maaroufa Press.

McGee, T. G. 1981. Labour mobility in fragmented labour markets, rural–urban linkages, and regional development in Asia. In *Rural–urban relations and regional development,* Fu-Chen Lo (ed.), 245–63. Nagoya: Maruzen Asia, for UN Centre for Regional Development.

Meadows, D. H., D. L. Meadows, J. Randers, & W. W. Behrens III 1972. *The limits to growth: a report for the Club of Rome's project on the predicament of mankind.* New York: Universe.

Mehmet, O. 1978. *Economic planning and social justice in developing countries.* New York: St Martin's Press.

Meinig, D. W. 1962. A comparative historical geography of two railnets: Columbia Basin and South Australia. *Annals, Association of American Geographers* 52, 394–413.

Meinig, D. W. 1969. A macrogeography of Western imperialism: some morphologies of moving frontiers of political control. In *Settlement and encounter: geographical studies presented to Sir Grenfell Price,* F. Gale & G.H. Lawton (eds), 213–40. Melbourne: Oxford University Press.

Mellor, R. E. H. 1975. *Eastern Europe: a geography of the Comecon Countries.* London: Macmillan.

Meyer, D. R. 1983. Emergence of the American manufacturing belt: an interpretation. *Journal of Historical Geography* 9, 145–74.

Micklin, P. P. 1986. The status of the Soviet Union's north–south water transfer projects before their abandonment in 1985–86. *Soviet Geography* 27, 287–329.

Micklin, P. P. 1988. Desiccation of the Aral Sea: a water management disaster in the Soviet Union. *Science* 241, 1170–6.

Mikesell, R. F. 1979. *New Patterns of world mineral development.* Washington DC: National Planning Association, for the British – North American Committee.

Milewski, J. J. 1981. Capitalism in Nigeria and problems of dependence: some historical comments. In *Dependency theory: a critical reassessment,* D. Seers (ed.), 109–18. London: Frances Pinter.

Miller, S. M. 1975. Notes on neo-capitalism. *Theory and Society* 2, 1–35.

Mohl, R. A. & N. Betten (eds) 1970. *Urban America in historical perspective.* New York: Weybright and Talley.

Mokyr, J. 1985. The Industrial Revolution and the new economic history. In *The economics of the Industrial Revolution,* J. Mokyr (ed.), 1–52. Totowa, NJ: Rowan and Allenheld.

Mollenkopf, J. H. 1981. Paths towards the post industrial service city: the northeast and the southwest. In *Cities under stress: the fiscal crises of urban America,* R. W. Burchell & D. Listokin (eds), 77–112. New Brunswick, NJ: Center for Urban Policy Research, Rutgers University.

Morgan, K. & A. Sayer 1985 A 'modern' industry in a 'mature' region: the remaking of management–labour relations. *International Journal of Urban and Regional Research* 9, 383–404.

Morris, R. S. 1978. *Bumrap on America's cities.* Englewood Cliffs, NJ: Prentice–Hall.

Mouly, J. 1979. Employment: a concept in need of renovation. In *Employment: outlook and insights,* D. H. Freedman (ed.), 111–17. Geneva: ILO.

Muller, P. O. 1973. Trend surfaces of American agricultural patterns: a macro–Thünian analysis. *Economic Geography* 49, 228–42.

Muller, R. E. 1980. *Revitalizing America.* New York: Simon and Schuster.

Murphey, R. 1980. *The fading of the Maoist vision: city and country in China's development.* New York: Methuen.

Murray, P. & I. Szelenyi 1984. The city in the transition to socialism. *International Journal of Urban and Regional Research* 8, 90–107.

Myrdal, A. 1978. The game of disarmament. In *Disarmament and world development,* R. Jolly (ed.), 85–92. Oxford: Pergamon Press.

Myrdal, G. 1957. *Economic theory and underdeveloped regions.* London: Duckworth.

Myrdal, G. 1970. The 'soft state' in underdeveloped countries. In *Unfashionable economics: essays in honour of Lord Balogh,* P. Streeten (ed.), 227–43. London: Weidenfeld and Nicolson.

Nafziger, E. W. 1976. A critique of development economics in the U.S. *Journal of Development Studies* 13, 18–34.

Nafziger, E. W. 1983. *The economics of political instability: the Nigerian–Biafran war.* Boulder, Col.: Westview.

Nef, J. U. 1964. *The conquest of the material world.* Chicago: University of Chicago Press.

Negandhi, A. R. & B. R. Baliga 1981. *Tables are turning: German and Japansese multinationals in the United States*. Cambridge, Mass.: Oelgeschlager, Gunn & Hain.

Nelson, J. M. 1969. *Migrants, urban poverty, and instability in developing countries*. Occasional Papers in International Affairs 22. Harvard University, Center for International Affairs.

Norcliffe, G. B. 1982. Informal industry and development observations based on the operating characteristics of enterprises in Central Province, Kenya. In *Industrial decline and regeneration: proceedings of the 1981 Anglo-Canadian Symposium*, L. Collins (ed.), 237–47. Edinburgh: Department of Geography and the Centre of Canadian Studies, University of Edinburgh.

North, D. C. 1955. Location theory and regional economic growth. *Journal of Political Economy* 62, 243–58.

North, D. C. & R. P. Thomas 1973. *The rise of the Western world: a new economic history*. Cambridge: Cambridge University Press.

Norton, R. D. 1979. *City life-cycles and American urban policy*. New York: Academic Press.

Norton, R. D. 1987. The once and present urban crisis. *Urban Studies* 24, 480–8.

Norton, R. D. & J. Rees 1979. The product cycle and the spatial decentralization of American manufacturing. *Regional Studies* 13, 141–51.

Nove, A. 1969. *An economic history of the U.S.S.R.* London: Penguin.

Nove, A. 1977. *The Soviet economic system*, London: Allen & Unwin.

Nove, A. 1979. Recent developments in East European economics. In *Post-industrial society*, B. Gustafsson (ed.), 119–42. London: Croom Helm.

Obregon, A. Q. 1974. The marginal pole of the economy and the marginalised labour force. *Economy and Society* 3, 393–428.

O'Brien, D. P. 1975. *The classical economists*. Oxford: Oxford University Press.

O'Brien, P. 1982. European economic development: the contribution of the periphery. *Economic History Review* 35, 1–18.

O'Connor, J. R. 1973. *The fiscal crisis of the state*. New York: St Martin's Press.

Odell, P. R. 1981. *Oil and world power*, 6th edn. London: Penguin.

Odell, P. R. & D. A. Preston 1973. *Economies and societies in Latin America: a geographical interpretation*. London: Wiley.

OECD 1983. *OECD employment outlook*. Paris: OECD.

Openshaw, S. & P. Steadman 1983a. Predicting the consequences of a nuclear attack on Britain: models, results, and implications for public policy. *Environment and Planning C: Government and Policy* 1, 205–28.

Openshaw, S. & P. Steadman 1983b. The geography of two hypothetical nuclear attacks on Britain. *Area* 15, 193–201.

Openshaw, S., P. Steadman & O. Greene 1983. *Doomsday: Britain after nuclear attack*. Oxford: Basil Blackwell.

Pacione, M. 1984. *Rural geography*. London: Harper & Row.

Pahl, R. E. 1980. Employment, work and the domestic division of labour. *International Journal of Urban and Regional Research* 4, 1–20.

Palma, G. 1981. Dependency and development: a critical overview. In *Dependency theory, a critical reassessment*, D. Seers (ed.), 20–78. London: Frances Pinter.

Parker, W. H. 1968. *An historical geography of Russia*. London: University of London Press.

Parker, W. H. 1972. *The superpowers: the United States and the Soviet Union compared*. London: Macmillan.

Pazos, F. 1973. Regional integration of trade among less developed countries. *World Development* 1(7), 1–12.

Peattie, L. R. 1980. Anthropological perspectives on the concepts of dualism, the informal sector, and marginality in developing urban economies. *International Regional Science Review* 5, 1–31.

Peet, R. 1969. The spatial expansion of commercial agriculture in the nineteenth century: a von Thünen interpretation. *Economic Geography* 45, 283–301.

Peet, R. 1983. Relations of production and the relocation of United States manufacturing industry since 1960. *Economic Geography* 59, 112–43.

Perez, C. 1985. Long waves and changes in socioeconomic organisation. *IDS Bulletin*, Sussex 16(1), 31–5.

Perry, D. C. & A. J. Watkins 1977. People, profit and the rise of the Sunbelt cities. In *The rise of the Sunbelt cities*, D. C. Perry & A. J. Watkins (eds), 277–307. Beverley Hills: Sage.

Polanyi, K. 1968. *Primitive, archaic and modern economies: essays of Karl Polanyi*, Boston: Beacon Press.

Pred, A. R. 1966. *The spatial dynamics of U.S. urban–industrial growth, 1800–1914*. Cambridge, Mass.: MIT Press.

Pred, A. R. 1973. *Urban growth and the circulation of information: The United States system of cities, 1790–1840*. Cambridge, Mass.: Harvard University Press.

Pred, A. 1977. *City-systems in advanced economies: past growth, present processes and future development options*. London: Hutchinson.

Price, K. A. (ed.) 1982. *Regional conflict and national policy*. Baltimore: Johns Hopkins University Press, for Resources for the Future, Inc.

Quante, W. 1976. *The exodus of corporate headquarters from New York City*. New York: Praeger.

Ray, D. M. 1971a. *Dimensions of Canadian Regionalism*. Geographical Paper No. 49. Ottawa: Department of Energy, Mines and Resources.

Ray, D. M. 1971b. The location of United States manufacturing subsidiaries in Canada. *Economic Geography* 47, 389–400.

Reed, H. C. 1983. International financial center preeminence in MNC and nation–state interaction. In *Governments and Multinationals*, W. Goldberg (ed.), 128–58. New York: Oelgeschlager, Gunn & Hain.

Reed, M. C. 1969. Railways and the growth of the capital market. In *Railways in the Victorian economy: studies in finance and economic growth*, M. C. Reed (ed.), 162–83. Newton Abbot: David and Charles.

Rees, J. 1979. Technological change and regional shifts in American manufacturing. *Professional Geographer* 31, 45–54.

Regional Studies Association 1983. *Report of an inquiry into regional problems in the United Kingdom*. Norwich: Geo Books.

Renshaw, G. 1981. *Employment, trade and North–South co-operation*. Geneva: ILO.

Revelle, R. 1982. Resources. In *Population and the world economy in the 21st century*, J. Faaland (ed.), 50–77. Oxford: Basil Blackwell.

Ricardo, D. 1817/1911. *The principles of political economy and taxation*. London: Dent.

Rodgers, A. 1974. The locational dynamics of Soviet industry. *Annals, Association of American Geographers* 64, 226–40.

Rogerson, C. M. 1980. Internal colonialism, transnationalization and spatial inequality. *South African Geographical Journal* 62, 103–20.

Rostankowski, P. 1980. The Nonchernozem Development Program and prospective spatial shifts in grain production in the agricultural triangle of the Soviet Union. *Soviet Geography* 21, 409–18.

Rostow, W. W. 1960. *The stages of economic growth: a non-communist manifesto.* Cambridge: Cambridge University Press.

Rostow, W. W. 1978. *The world economy: history and prospect.* Austin: University of Texas Press.

Rothwell, R. 1982. The role of technology in industrial change: implications for regional policy. *Regional Studies* 16, 361–9.

Rycroft, R. W. & J. E. Monaghan 1982. National security policy: synfuels and the MX system. In *Energy and the western United States: politics and development.* J. L. Regens, R. W. Rycroft & G. A. Daneke (eds), 69–100. New York: Praeger.

Sachs, I. 1976. *The discovery of the Third World.* Cambridge, Mass.: MIT Press.

Sagers, M. J. & M. B. Green 1979. Industrial dispersion in the Soviet Union: An application of entropy measures. *Soviet Geography* 20, 567–85.

Sampson, A. 1977. *The arms bazaar: the companies, the dealers, the bribes: from Vickers to Lockheed,* London: Hodder.

Sandbrook, R. 1986. The state and economic stagnation in tropical Africa. *World Development,* 14, 319–32.

Santos, M. 1977. Spatial dialectics; the two circuits of urban economy in underdeveloped countries. *Antipode* 9(3), 49–60.

Saul, S. B. 1960. *Studies in British Overseas Trade 1870–1914.* Liverpool: Liverpool University Press.

Saul, S. B. 1972. The nature and diffusion of technology. In *Economic Development in the long run,* A. J. Youngson (ed.), 36–61. London: Allen & Unwin.

Sayer, A. 1986. Industrial location on a world scale: the case of the semiconductor industry. In *Production, work, territory,* A. J. Scott & M. Storper (eds.) 107–23. Boston: Allen & Unwin.

Schiffer, J. R. 1985. Interpretations of the issue of 'inequality' in Soviet regional policy debates. *International Journal of Urban and Regional Research* 9, 508–32.

Schoenberger, E. 1988. From fordism to flexible accumulation: technologies, competitive strategies and international location. *Society and Space* 6, 245–62.

Science Council of Canada 1980. *Multinationals and industrial strategy: the role of the world product mandates.* Ottawa: Science Council of Canada.

Science Council of Canada 1986. *A growing concern: soil degradation in Canada.* Ottawa: Science Council of Canada.

Scott, A., J. Olynyk & S. Renzetti 1986. The design of water-export policy. In *Canada's resource industries and water export policy,* J. Whalley (coord), 161–246 Toronto: University of Toronto Press.

Scott, A. J. 1982. Production system dynamics and metropolitan development. *Annals, Association of American Geographers* 72, 185–200.

Scott, A. J. 1986. Industrial organisation and location: division of labour, the firm and spatial process. *Economic Geography* 62, 215–31.

Seers, D. 1970. Graduate migration as an obstacle to equality. In *Unfashionable economics: essays in honour of Lord Balogh,* P. Streeten (ed.),

98–107. London: Weidenfeld and Nicolson.

Seers, D. 1979a. Introduction. In *Underdeveloped Europe: studies in core-periphery relations*, D. Seers, B. Schaffer & M.-L. Kiljunen (eds), xiii–xxi. Hassocks, Sussex: Harvester Press.

Seers, D. 1979b. The congruence of Marxism and other neoclassical doctrines. In *Toward a new strategy for development*, Rothko Chapel Colloquium, 1–17. New York: Pergamon Press.

Seers, D. 1981. Development options: the strengths and weaknesses of dependency theories in explaining a government's room to manoeuvre. In *Dependency theory: a critical reassessment*, D. Seers (ed.), 135–49. London: Frances Pinter.

Segal, N. S. 1979. The limits and means of 'self-reliant' regional economic growth. In *Regional policy: past experience and new directions*, D. Maclellan & J. B. Parr (eds), 212–24. Oxford: Martin Robertson.

Sella, D. 1974. European industries, 1500–1700. *In The Fontana economic history of Europe*, Vol. 2: *The Sixteenth and seventeenth centuries*, C. M. Cipolla (ed.), 354–426. Glasgow: Fontana.

Selucky, R. 1979. *Marxism, socialism, freedom: towards a general democratic theory of labour-managed systems*. New York: St Martin's Press.

Shabad, T. 1978. Some geographic aspects of the new Soviet Five-Year Plan. *Soviet Geography* 19, 202–5.

Shabad, T. 1986. Geographic aspects of the new Soviet Five-Year Plant. *Soviet Geography* 27, 1–16.

Singh, M. S. & C. S. Choo 1981. Spatial dynamics in the growth and development of multinational corporations in Malaysia. In *Spatial analysis, industry and the industrial environment: progress in research and applications*, Vol. II: *International industrial systems*, F. E. I. Hamilton & G. J. R. Linge (eds), 481–507. Chichester: Wiley.

Sivard, R. L. 1982. *World military and social expenditures 1982*. Leesburg, Va.: World Priorities.

Sivard, R. L. 1986. *World military and social expenditures 1986*. Leesburg, Va.: World Priorities.

Slater, D. 1982. State and territory in postrevolutionary Cuba: some critical reflections on the development of spatial policy. *International Journal of Urban and Regional Research* 6, 1–34.

Smith, C A. 1976. Exchange systems and the spatial distribution of elites: the organization of stratification in agrarian societies. In *Regional analysis*, Vol. II: *Social systems*, C. A. Smith (ed.), 309–74. New York: Academic Press.

Smith, N. 1982. Gentrification and uneven development. *Economic Geography* 58, 139–55.

Smith, W. 1984. The 'vortex model' and the changing agricultural landscape of Quebec. *Canadian Geographer* 28, 358–72.

Soja, E., R. Morales & G. Wolff 1983. Urban restructuring: an analysis of social and spatial change in Los Angeles. *Economic Geography* 59, 195–230.

Spooner, B. & R. Netting 1972. Humanised economics. *Peasant Studies Newsletter* 1, 54–8.

Stanback, T. M., P. J. Bearse, T. J. Noyelle & R. A. Karasek 1981. *Services: the new economy*. Totowa, NJ: Rowman and Allanheld.

Stavrianos, L. S. 1981. *Global rift: the Third World comes to age*. New York: Morrow.

Steed, G. P. F. 1976. Locational factors and dynamics of Montreal's large garment complex. *Tijdschrift voor Economische en Sociale Geografie* 67, 151–68.

Steed, G. P. F. 1981. International location and comparative advantage: the clothing industries and developing countries. In *Spatial analysis, industry and the industrial environment: progress in research and applications*, Vol. II: *International industrial systems*, F. E. I. Hamilton & G. J. R. Linge (eds), 265–303. Chichester: Wiley.

Steed, G. P. F. & D. DeGenova 1983 Ottawa's technology-oriented complex. *Canadian Geographer* 27, 263–78.

Stein, J. M. 1985. Militarism as a domestic planning issue. *International Journal of Urban and Regional Research* 9, 341–51.

Stern, R. M. 1985. Global dimensions and determinants of international trade and investment in services. In *Trade and investment in services*, Ontario Economic Council, 126–68. Toronto: Ontario Economic Council.

Sternlieb, G. & J. W. Hughes (eds), 1975. *Post-industrial America: metropolitan decline and inter-regional job shifts*. New Brunswick, NJ: Center for Urban Policy Research, Rutgers University.

Stewart, F. 1976. The direction of international trade: gains and losses for the Third World. In *A world divided: the less developed countries in the international economy*, G. K. Helleiner (ed.), 89–110 Cambridge: Cambridge University Press.

Stohr, W. B. & D. R. F. Taylor 1981. Development from above or below? Some conclusions. In *Development from above or below? The dialectics of regional planning in developing countries*, W. B. Stohr & D. R. F. Taylor (eds), 453–80. Chichester: Wiley.

Stopford, J. M. 1976. Changing perspectives on investment by British manufacturing multinationals. *Journal of International Business Studies* 7, (2), 15–27.

Supple, B. 1972. Thinking about economic development. In *Economic development in the long run*, A. J. Yongson (ed.), 19–35. London: Allen & Unwin.

Susman, P., P. O'Keefe & B. Wisner (1983) Global disasters, a radical interpretation. In *Interpretations of calamity*, K. Hewitt (ed.), 263–83. Boston: Allen & Unwin.

Symons, L. 1972. *Russian agriculture: a geographic survey*. London: Bell.

Taaffe, E. J., R. L. Morrill & P. R. Gould 1963. Transport expansion in under-developed countries. *Geographical Review* 53, 503–29.

Tabb, W. K. 1978. The New York City fiscal crisis. In *Marxism and the metropolis: new perspectives in urban political economy*, W. K. Tabb & L. Sawers, (eds), 241–66. New York: Oxford University Press.

Tarrant, J. R. 1980. The geography of food aid. *Transactions, Institute of British Geographers, NS* 5, 125–40.

Taylor, M. 1986. The product-cycle model: a critique. *Environment and planning A* 18, 751–61.

Thompson, E. P. 1968. *The making of the English working class*. London: Penguin.

Thrift, N. 1979. Unemployment in the inner city: urban problem or structural imperative? A review of the British experience. In *Geography and the urban environment: progress in research and applications*, Vol. II, D. T. Herbert & R. J. Johnston (eds), 125–226. Chichester: Wiley.

Thrift, N. 1985. Research policy and review 1. Taking the rest of the world seriously? The state of British urban and regional research in a time of economic crisis. *Environment and Planning A* 17, 7–24.

Thrift, N. 1986. The geography of international economic disorder. In *A world in crisis? Geographical perspectives*, R. J. Johnston & P. J. Taylor, (eds), 12–67, Oxford: Basil Blackwell.

Todaro, M. P. 1977. *Economic development in the Third World: an introduction to problems and policies in a global perspective*. New York: Longman.

Townsend, A. R. 1977. The relationship of inner city problems to regional policy. *Regional studies* 11, 225–51.

Townsend, A. R. 1983. *The impact of recession – on industry, employment and the regions, 1976–1981*. London: Croom Helm.

Troughton, M. 1982. Process and response in the industrialisation of agriculture. In *The effect of modern agriculture on rural development*, G. Enyedi & I. Volgyes, (eds.), 213–27. New York: Pergamon Press.

Tulchinsky, G. J. J. 1977. *The river barons: Montreal businessmen and the growth of industry and transportation 1837–53*. Toronto: University of Toronto Press.

Turner, L. 1982a. Consumer electronics: the colour television case. In *The newly industrializing countries: trade and adjustment*, L. Turner & N. McMullen (eds), 48–68. London: Allen & Unwin.

Turner, L. 1982b. The view from the NICs. In *The newly industrializing countries: trade and adjustment*, L. Turner & N. McMullen (eds), 160–8. London: Allen & Unwin.

Turnock, D. 1978. *Eastern Europe: studies in industrial geography*. Folkestone: Dawson.

UNIDO 1979. *World industry since 1960: progress and prospects*. Vienna: United Nations Industrial Development Organization.

UNIDO 1981. *World Industry in 1980*. Vienna: United Nations Industrial Development Organization.

UN (United Nations) 1978. *Transnational corporations in world development: a re-examiantion*. New York: UNESCO.

UN 1983a. *Salient features and trends in foreign direct investment*. New York: UN Centre on Transnational Corporations.

UN 1983b. *Transnational corporations in world development: third survey*. New York: UN Centre on Transnational Corporations.

Uphoff, N. & M. J. Esman 1978. *Local organization for rural development: analysis of Asian experience*. Ithaca, NY: Cornell University.

US Congress 1986. *Technology, public policy, and the changing structure of American agriculture*. Washington, DC: Office of Technology Assessment, US Congress.

US Government 1980/2. *The Global 2000 report to the President: entering the twenty-first century*. New York: Penguin.

Vance, J. E. 1970. *The merchant's world: the geography of wholesaling*. Englewood Cliffs, NJ: Prentice-Hall.

Vance, J. E. 1977. *This scene of man: the role and structure of the city in the geography of Western civilization*. New York: Harper & Row.

Vernon, R. 1966 International investment and international trade in the product cycle. *Quarterly Journal of Economics* 80, 190–207.

Vernon, R. 1979. The product cycle hypothesis in a new international environment. *Oxford Bulletin of Economics and Statistics* 41, 255–68.

Viljoen, S. 1974. *Economic systems in world history*. London: Longman.

Vining, D. R., J. B. Dobronyi, M. A. Otness & J. E. Schwinn 1982. A principal axis shift in the American spatial economy. *Professional Geographer* 34, 270–8.

Vogeler, I. 1981. *The myth of the family farm: agribusiness dominance of US agriculture*. Boulder, Col.: Westview.

Von Thünen, H. J. 1826. *Der Isolierte Staat. Von Thünen's Isolated state*, P. Hall (ed.) Oxford: Pergamon Press (English translation 1966).

Waddell, E. W. 1972. *The mound builders: agricultural practices, environment and society in the Central Highlands of New Guinea*. Seattle: University of Washington Press.

Walker, D. F. 1980. *Canada's industrial space-economy*. Toronto: Wiley.

Wallace, I. 1978. Towards a humanized conception of economic geography. In *Humanistic geography: prospects and problems*, D. Ley & M. S. Samuels (eds), 91–108. Chichago: Maaroufa Press.

Wallace, I. 1985. Towards a geography of agribusiness. *Progress in Human Geography* 9, 491–514.

Wallerstein, I. 1976. *The modern world-system: capitalist agriculture and the origins of the European world-economy in the sixteenth century*. New York: Academic Press.

Wallerstein, I. 1980. *The modern world-system II: mercantilism and the consolidation of the European world-economy, 1600–1750*. New York: Academic Press.

Ward, B. 1976. *The home of man*. London: André Deutsch.

Warren, B. 1979. The postwar economic experience of the Third World. In *Toward a new strategy for development*, Rothko Chapel Colloquium, 144–68. New York: Pergamon Press.

Warrick, R. A. 1983. Drought in the U.S. Great Plains: shifting social consequences? In *Interpretations of calamity*, K. Hewitt (ed.), 67–82. Boston: Allen & Unwin.

Watkins, M. H. 1963. A staple theory of economic growth. *Canadian Journal of Economics and Political Science* 29, 141–58.

Watts, H. D. 1979. Large firms, multinationals and regional development: some new evidence from the United Kingdom. *Environment and Planning A* 11, 71–81.

Weber, A. 1909. *Alfred Weber's theory of the location of industries*. Chicago: University of Chicago Press (English translation 1929).

Weinstein, F. B. 1976. Multinational corporations and the Third World: the case of Japan and South East Asia. *International Organization* 30, 373–404.

Weller, G. R. 1977. Hinterland politics: the case of Northwestern Ontario. *Canadian Journal of Political Science* 10, 727–54.

Wells, L. T. 1972. *The product life cycle and international trade*. Cambridge, Mass.: Harvard University Press.

Wessel, J. 1983. *Trading the future: farm exports and the concentration of economic power in our food system*, San Francisco: Institute for Food Development Policy.

Wheatley, P. 1971. *The pivot of the four quarters: a preliminary enquiry into the origins and character of the ancient Chinese city*. Chicago: Aldine.

Wilkinson, R. G. 1973. *Poverty and progress: an ecological model of economic development*. London: Methuen.

Williamson, J. C. 1965. Regional inequality and the process of national development: a description of the patterns. *Economic Development and Cultural Change* 13, 3–45.

Wolch, J. R. 1983. The voluntary sector in urban communities. *Society and Space* 1, 181–90.

Wolfe, R. I. 1968. Economic development. In *Canada: a geographical interpretation*, J. Warkentin (ed.), 187–228. Toronto: Methuen.

Woolcock, S. 1982. Textiles and clothing. In *The newly industrializing countries: trade and adjustment*, L. Turner & N. McMullen (eds), 27–47. London: Allen & Unwin.

World Bank 1981. *World development report 1981*. New York: Oxford University Press.

World Bank 1982. *World development report 1982*. New York: Oxford University Press.

World Bank 1983. *World development report 1983*. New York: Oxford University Press.

World Bank 1986. *World development report 1986*. New York: Oxford University Press.

World Bank 1988. *World development report 1988*. New York: Oxford University Press.

World Commission on Environment and Development (Bruntland Commission) 1987. *Our common future*. New York: Oxford University Press.

World Resources Institute 1986. *World resources 1986*. New York: Basic Books.

Wrigley, C. C. 1978. Neo-mercantile policies and the new imperialism. In *The imperial impact: studies in the economic history of Africa and India*. C. Dewey & A. G. Hopkins (eds), 20–34. London: Athlone Press.

Wrigley, E. A. 1978. A simple model of London's importance in changing English society and economy 1650–1750. In *Towns in societies: essays in economic history and historical sociology*, P. Abrams & E. A. Wrigley (eds), 215–43. Cambridge: Cambridge University Press.

Wynn, G. 1987. The maritimes: the geography of fragmentation and underdevelopment. In *Heartland and hinterland: a geography of Canada*, 2nd edn, L. D. McCann (ed.), 175–245. Toronto: Prentice-Hall Canada.

Youngson, A. J. 1965. The opening up of new territories. In *The Cambridge economic history of Europe*, Vol. VI: *The Industrial Revolutions and after: incomes, population and technological change (1)*, H. J. Habakkuk & M. Postan (eds), 139–211. Cambridge: Cambridge University Press.

Index